Key Drivers and Indicators of Soil Health Management

Ajay Kumar Mishra •
Sheetal Sharma • Antaryami Mishra
Editors

Key Drivers and Indicators of Soil Health Management

Transitioning from Conventional to Regenerative Agriculture

Springer

Editors
Ajay Kumar Mishra
South Asia Regional Centre
International Rice Research Institute
Varanasi, Uttar Pradesh, India

Sheetal Sharma
International Rice Research Institute
New Delhi, Delhi, India

Antaryami Mishra
Odisha University of Agriculture & Technology
Bhubaneswar, Odisha, India

ISBN 978-981-97-7563-7 ISBN 978-981-97-7564-4 (eBook)
https://doi.org/10.1007/978-981-97-7564-4

This Springer imprint is published by the registered company Springer Nature Singapore Pte Ltd.
The registered company address is: 152 Beach Road, #21-01/04 Gateway East, Singapore 189721, Singapore

Foreword

भारत सरकार
कृषि अनुसंधान और शिक्षा विभाग एवं
भारतीय कृषि अनुसंधान परिषद
कृषि एवं किसान कल्याण मंत्रालय, कृषि भवन, नई दिल्ली–110 001

सत्यमेव जयते

डॉ. हिमांशु पाठक
DR. HIMANSHU PATHAK
सचिव (डेयर) एवं महानिदेशक (आईसीएआर)
Secretary (DARE) &
Director General (ICAR)

GOVERNMENT OF INDIA
DEPARTMENT OF AGRICULTURAL RESEARCH AND EDUCATION (DARE)
AND
INDIAN COUNCIL OF AGRICULTURAL RESEARCH (ICAR)
MINISTRY OF AGRICULTURE AND FARMERS WELFARE
Krishi Bhavan, New Delhi 110 001
Tel: 23382629 / 23386711 Fax: 91-11-23384773
E-mail: dg.icar@nic.in

Regenerative agriculture is a promising approach to farming that aims to restore and enhance the health of ecosystems through sustainable practices. It focuses on improving soil health, increasing biodiversity, and enhancing ecosystem services. Its holistic and adaptive strategies offer pathways to more sustainable and resilient food systems. Simultaneously, the challenges with respect to initial transition costs, learning curves, understanding ecosystem dynamics, market access, and policy issues need to be analyzed. This book, an essential compilation on this subject, provides a holistic examination and actionable insights to transform agricultural production systems to support food production and ecological resilience.

I commend the editors and contributors for their meticulous efforts in curating a range of topics that span the practical applications of regenerative practices, offering a blueprint for future agricultural strategies that align with environmental conservation and enhanced food security. The chapters serve as a testament to the innovative spirit of our agricultural community, striving towards systems that restore, rather than deplete, our precious natural resources. I congratulate the editors and authors for bringing out this book, which I hope will bridge the knowledge gaps in various aspects of regenerative agriculture research. The insights from this book will undoubtedly contribute to shaping the policies and practices that guide sustainable agriculture across the globe. I compliment editors and authors for their valuable contributions in this book and hope that the book will serve as an excellent source of reference material to researchers, farmers, students, and other stakeholders.

New Delhi, Delhi, India

10th July, 2024 — Himanshu Pathak

Preface

The dawn of the twenty-first century has heralded unprecedented challenges and opportunities in the realm of agriculture. As global populations burgeon, the demand for food security, environmental sustainability, and economic viability becomes more pressing. In response to these multifaceted demands, the concept of regenerative agriculture has emerged as a beacon of hope and innovation. This book, "Key Drivers and Indicators of Soil Health Management: Transitioning from Conventional to Regenerative Agriculture," is a comprehensive exploration of the principles, practices, and impacts of regenerative agriculture on various aspects of the ecosystem and society.

Regenerative agriculture is more than a collection of farming techniques; it represents a paradigm shift towards holistic land management that prioritizes soil health, biodiversity, and well-being of farming communities. The chapters within this book delve into diverse aspects of regenerative agriculture, from soil health assessments and microbial community dynamics to the integration of digital innovations in agro-advisory services and the optimization of nutrient management.

The contributors to this book are esteemed experts from various fields, including soil science, agronomy, environmental sciences, and digital agriculture. Their collective expertise and insights provide a rich tapestry of knowledge that underscores the transformative potential of regenerative practices.

Readers will gain insights into the specific challenges and solutions pertinent to soil management in different regions. The crucial role of microbial communities in maintaining soil health and productivity is also explored. The book highlights the intersection of technology and traditional farming practices, demonstrating how modern tools can enhance the efficacy and reach of regenerative agriculture. Additionally, practical guidelines for addressing critical resource management issues are provided.

This book is intended for a wide audience, including researchers, practitioners, policymakers, and students. It aims to bridge the gap between scientific research and practical application, offering a valuable resource for anyone interested in fostering sustainable agricultural practices.

We extend our deepest gratitude to all the contributors and reviewers who have made this book possible. Their dedication and expertise are the bedrock upon which this work stands. We also acknowledge the support of our respective institutions and organizations, whose commitment to advancing agricultural sustainability has been unwavering.

As you embark on this journey through the pages of "Regenerative Agriculture: Strategies and Practices for Sustainable Farming," we hope that you find inspiration, knowledge, and a renewed commitment to cultivating a healthier, more resilient world.

Varanasi, Uttar Pradesh, India Ajay Kumar Mishra
New Delhi, Delhi, India Sheetal Sharma
Bhubaneswar, Odisha, India Antaryami Mishra

Acknowledgments

We want to acknowledge the support of our respective institutions and organizations, whose commitment to advancing agricultural sustainability has been unwavering. This book, "Key Drivers and Indicators of Soil Health Management: Transitioning from Conventional to Regenerative Agriculture," would not have been possible without their continuous encouragement and resources.

Special thanks to Dr. Sudhanshu Singh, Director, IRRI South Asia Regional Centre, for his encouragement and for his strategic directions and innovative approach to regenerative agriculture, which have significantly shaped our work and this book. Prof. Shinya Funakawa and Prof. Hitoshi Shinjo from Kyoto University are also acknowledged for their invaluable insights, particularly their pioneering research on sustainable land management practices and soil health, which have deeply influenced our perspectives and inspired us to explore new frontiers in regenerative agriculture. Additionally, we are grateful to Dr. Suresh Kumar Chaudhari, Deputy Director General (Natural Resource Management), ICAR Headquarters, New Delhi, for his visionary leadership in promoting integrated nutrient management and sustainable agricultural practices across India. His dedication to advancing soil and water conservation techniques has been a cornerstone in our understanding and advocacy for natural resource management.

We also extend our sincere thanks to Dr. Parveen Kumar, Director, ICAR-CCARI, Goa, for his relentless efforts in advancing climate-resilient farming systems and agroecological research. His work has provided a strong foundation for our exploration of diversified cropping systems and their role in enhancing farm resilience. Dr. ML Jat, Global Research Program Director, Resilient Farm and Food Systems at ICRISAT, has been instrumental in shaping our approach to resilient farming practices through his extensive research on climate-smart agriculture and resource-conserving technologies. Dr. HS Jat, Director, ICAR-Indian Institute of Maize Research, has significantly contributed to our understanding of sustainable maize-based systems and integrated pest management, enriching our knowledge base and inspiring us to incorporate these principles into our writing.

I would like to express my heartfelt appreciation to my family for their firm support and inspiration: my late grandfather, Shri Keshav Mishra; my father, Shri Chandra Bhushan Mishra; my uncle, Dr. Ram Bhushan Mishra; my mother, Smt. Kamalawati Mishra; my father-in-law, late Shri Ramanand Mishra; my wife, Ms.

Sarita Mishra; and my son, Atharv Mishra. Their love and encouragement have been the cornerstone of my journey.

As you embark on this journey through the pages of "Transition to Regenerative Agriculture: Principles and Indicators of Soil Health Management," we hope that you find inspiration, knowledge, and a renewed commitment to cultivating a healthier, more resilient world.

Contents

Editors and Contributors

About the Editors

Ajay Kumar Mishra is a highly accomplished senior associate scientist in the field of soil science, currently working with the International Rice Research Institute (IRRI). With expertise in enhancing rice-based agri-food systems through improved soil nutrient management and agri-relevant ICT, Mishra is at the forefront of strategic and adaptive research. Mishra's primary focus lies on the development and implementation of the Rice Crop Manager tool, an ICT-based platform that delivers context-specific rice farming management technologies. Collaborating with public and private partners, he actively contributes to the rapid deployment of actionable and tailored farming solutions. His work is particularly targeted towards small-scale farmers in India, where the application of these technologies helps improve profitability and productivity in rice and rice-based agri-food farming. Apart from his involvement in ICT-based solutions, Mishra is dedicated to developing low-cost innovative technologies for climate-smart agriculture and ensuring their adoption by the farming community. He has also played a pivotal role in the capacity development of extension agents and scholars, imparting knowledge and skills to drive positive change in agricultural practices. Mishra's research contributions are significant, as evident from his authorship and co-authorship of over 30 scientific papers in peer-reviewed journals and book chapters. His expertise, dedication, and collaborative approach make him an invaluable asset in advancing sustainable and efficient agricultural practices in the context of rice farming.

Sheetal Sharma is a highly accomplished senior scientist in the field of soil science, affiliated with the International Rice Research Institute (IRRI). With a strong background in agricultural research, Dr. Sharma has made significant contributions to the understanding of soil fertility, nutrient management, and sustainable rice production. Dr. Sharma holds a Ph.D. in soil science and has amassed over 15 years of experience in agricultural research and development. Her expertise lies in the areas of soil fertility, nutrient cycling, and impact of agricultural practices on soil health. She has conducted extensive research on the optimization of nutrient management strategies for rice production, focusing on improving crop productivity while minimizing environmental impacts. As a senior scientist at the IRRI, Dr. Sharma has been actively involved in conducting field experiments, analyzing soil samples, and providing valuable recommendations to farmers and stakeholders. She has collaborated with international research organizations, government agencies, and farmer communities to develop and promote sustainable agricultural practices in rice-growing regions. Dr. Sharma's research findings have been published in reputable scientific journals, and she has also contributed to various technical reports and publications. She has participated in conferences, workshops, and training programs to share her knowledge and expertise with fellow researchers and practitioners in the field. With her extensive research experience and dedication to sustainable rice production, Dr. Sheetal Sharma plays a crucial role in advancing soil science and promoting environmentally friendly agricultural practices. Her work at the IRRI contributes to global efforts in achieving food security and sustainable development in rice-based cropping systems.

Antaryami Mishra is a renowned expert in the field of soil science, affiliated with the Orissa University of Agriculture and Technology (OUAT) in Bhubaneswar, India. With a distinguished career spanning over two decades, Prof. Mishra has made significant contributions to the study and understanding of soil health management and sustainable agricultural practices. Prof. Mishra holds a Ph.D. in soil science and has established himself as a prominent researcher and academician in the field. He has authored numerous research papers, book chapters, and technical reports that have contributed to the advancement of knowledge in soil fertility, nutrient management, and soil health assessment. As a professor at OUAT, Prof. Mishra has been actively involved in teaching and mentoring students pursuing degrees in soil science and related fields. He has guided and supervised several research projects and has been instrumental in shaping the careers of aspiring soil scientists. Prof. Mishra's expertise and knowledge extend beyond the academic realm. He has collaborated with government agencies, NGOs, and farmer communities to promote sustainable agricultural practices and raise awareness about soil conservation, nutrient management, and sustainable land use. With his extensive research contributions and dedication to the field of soil science, Prof. Antaryami Mishra continues to inspire and guide future generations of soil scientists, making a significant impact on the agricultural landscape and sustainable development in India.

Contributors

Sumit Acharya Department of Horticulture, M. S. Swaminathan School of Agriculture (MSSSoA), Centurion University of Technology and Management (CUTM), Gajapati, Odisha, India

Protik Banerjee The University of Texas, San Antonio, TX, USA

Biswaranjan Baraj Sambodhi Research and Communications Pvt. Ltd., Noida, Uttar Pradesh, India

Saikumar C. Bharamappanavara Sambodhi Research and Communications Pvt. Ltd., Noida, Uttar Pradesh, India

Laliteshwari Bhardwaj Department of Botany, Institute of Science, Banaras Hindu University, Varanasi, Uttar Pradesh, India

Subhrajyoti Chatterjee Department of Horticulture, M. S. Swaminathan School of Agriculture (MSSSoA), Centurion University of Technology and Management (CUTM), Gajapati, Odisha, India

Pradeep Kumar Dash Varaha Climate Ag Pvt. Ltd, Gurugram, Haryana, India

Prava Kiran Dash Regional Research and Technology Transfer Station, Odisha University of Agriculture and Technology, Dhenkanal, India

Tarini Prasad Das Department of Agronomy, Odisha University of Agriculture and Technology, Bhubaneswar, Odisha, India

Nirmal De Department of Soil Science and Agricultural Chemistry, Institute of Agricultural Sciences, Banaras Hindu University, Varanasi, Uttar Pradesh, India

Krisanu Golui Department of Soil Science, Chaudhary Charan Singh Haryana Agricultural University, Hisar, Haryana, India

Rajman Gupta School of Environmental Sciences, Jawaharlal Nehru University, New Delhi, India

Chetan Gyawali Department of Soil Science and Agricultural Chemistry, Institute of Agricultural Sciences, Banaras Hindu University, Varanasi, Uttar Pradesh, India

Charchika Khanna Department of Soil Science and Agricultural Chemistry, Sam Higginbottom University of Agriculture and Technology, Prayagraj, Uttar Pradesh, India

Sabyasachi Koley Department of Soil Science and Agricultural Chemistry, Institute of Agricultural Sciences, Banaras Hindu University, Varanasi, Uttar Pradesh, India

Konathala Kusumavathi Department of Agronomy, Bidhan Chandra Krishi Viswavidyalaya, Mohanpur, Nadia, West Bengal, India

Indian Institute of Water Management, ICAR, Bhubaneswar, India

Athuman Juma Mahinda College of Agriculture and Food Technology, University of Dar es Salaam, Dar es Salaam, Tanzania

Piyush Kumar Maurya International Rice Research Institute, South Asia Regional Centre, Varanasi, Uttar Pradesh, India

Ajay Kumar Mishra International Rice Research Institute, South Asia Regional Centre, Varanasi, Uttar Pradesh, India

Antaryami Mishra Odisha University of Agriculture and Technology, Bhubaneswar, Odisha, India

Siksha O Anusandhan University, Bhubaneswar, Odisha, India

Sarita Mishra AgriPath, Grameen Foundation India Private Limited (GFI), Gurgaon, Haryana, India

Shraddha Mohanty Krishi Vigyan Kendra, Odisha University of Agriculture and Technology, Kalahandi, India

Kiran Kumar Mohapatra AICRP for Dryland Agriculture, Odisha University of Agriculture and Technology, Phulbani, India

Pritismita Nayak Department of Horticulture, M. S. Swaminathan School of Agriculture (MSSSoA), Centurion University of Technology and Management (CUTM), Gajapati, Odisha, India

Jimmy A. Ocaña-Reyes Estación Experimental Agraria Pucallpa, Dirección de Desarrollo Tecnológico Agrario, Instituto Nacional de Innovación Agraria (INIA), Pucallpa, Peru

Kshitipati Padhan Department of Agriculture and Allied Science, CV Raman Global University, Bhubaneswar, India

Narayan Panda Department of Soil Science and Agricultural Chemistry, Odisha University of Agriculture and Technology, Bhubaneswar, India

Girija Prasad Patnaik International Rice Research Institute, New Delhi, India

R. Arockia Infant Paul International Rice Research Institute, South Asia Regional Center, Varanasi, Uttar Pradesh, India

Pavithra Pidikiti International Rice Research Institute, South Asian Regional Centre, Varanasi, Uttar Pradesh, India

Amber Rai AgriPath, Grameen Foundation India Private Limited (GFI), Gurgaon, Haryana, India

H. S. Roopa AgriPath, Grameen Foundation India Private Limited (GFI), Gurgaon, Haryana, India

Sanjib Kumar Sahoo College of Agriculture, Odisha University of Agriculture and Technology, Bhawanipatna, India

Jhonsonraju Sankati International Rice Research Institute, South Asian Regional Centre, Varanasi, Uttar Pradesh, India

Udita Ghosh Sarkar Client Insights for Impact, Grameen Foundation India Private Limited (GFI), Gurgaon, Haryana, India

Debadatta Sethi Department of Soil Science and Agricultural Chemistry, Odisha University of Agriculture and Technology, Bhubaneswar, India

AICRP on PHET, College of Agricultural Engineering and Technology, Odisha University of Agriculture and Technology, Bhubaneswar, India

Sheetal Sharma International Rice Research Institute, South Asia Regional Centre, Varanasi, Uttar Pradesh, India

Yash Srivastava Sambodhi Research and Communications Pvt. Ltd., Noida, Uttar Pradesh, India

Prince Kumar Tiwari Department of Mechanical Engineering, Indian Institute of Technology BHU, Varanasi, Uttar Pradesh, India

Regenerative Agriculture: A Multifaceted Approach to One Health and Soil Restoration

1

Ajay Kumar Mishra, Pradeep Kumar Dash, Antaryami Mishra, and Sheetal Sharma

Introduction

Conventional agricultural (CA) practices have historically relied heavily on inorganic fertilizers, pesticides, and other agrochemicals to boost crop yields, leading to significant adverse impacts on ecosystems (Lal 2015a; Mishra et al. 2022). The detrimental effects of CA practices on ecosystems are crucial to explore, as tillage, a common practice, disrupts soil structure and organic matter, resulting in habitat loss for soil organisms and a decrease in microbial diversity (Lal 2006, 2015b). Soil structure is essential for supporting plant growth and water retention, and its degradation can lead to erosion and diminished soil fertility. Chemical fertilizers and pesticides further impact soil biodiversity by altering soil pH, degrading organic matter, and reducing microbial abundance and diversity (Brown et al. 2022; Bhaduri et al. 2022). Soil microbes play critical roles in nutrient cycling, soil health, and plant growth, and a decline in their diversity can destabilize ecosystems and compromise resilience (Giller et al. 2021; Mishra et al. 2024).

Moreover, the excessive use of agrochemicals can result in runoff, where these chemicals leach into water bodies, causing pollution that harms aquatic ecosystems by inducing algal blooms, depleting oxygen levels, and disrupting the balance of aquatic organisms. Additionally, pesticide residues can persist in water bodies,

A. K. Mishra (✉) · S. Sharma
International Rice Research Institute South Asia Regional Centre,
Varanasi, Uttar Pradesh, India
e-mail: a.k.mishra@irri.org

P. K. Dash
Varaha Climate Ag Pvt. Ltd., Gurugram, Haryana, India

A. Mishra
Odisha University of Agriculture and Technology, Bhubaneswar, Odisha, India

Siksha O Anusandhan University, Bhubaneswar, Odisha, India

A. K. Mishra et al. (eds.), *Key Drivers and Indicators of Soil Health Management*, https://doi.org/10.1007/978-981-97-7564-4_1

adversely affecting aquatic life and human health through bioaccumulation in the food chain. Burning crop residues, another widespread practice in conventional agriculture, releases harmful pollutants into the atmosphere, contributing to air pollution and posing risks to human health, wildlife, and the environment (Gordon et al. 2022). Additionally, burning crop residues depletes organic carbon and essential nutrients in the soil, exacerbating soil degradation. Monoculture, a hallmark of CA, reduces habitat diversity and disrupts ecosystems by diminishing the variety of plants and organisms present in agricultural landscapes, making them more susceptible to pests, diseases, and environmental stressors (Jordon et al. 2022; Jayasinghe et al. 2023). This loss of biodiversity has cascading effects on ecosystem functioning and resilience.

Consequently, CA practices, including intensive tillage and monoculture, contribute to soil erosion and land degradation, which strip away fertile topsoil, reduce land productivity, and exacerbate environmental issues like sedimentation in water bodies. Land degradation undermines the long-term sustainability of agricultural systems and jeopardizes their ability to support future generations (Kibblewhite et al. 2008; Khangura et al. 2023).

Regenerative agriculture (RA) offers an approach to farming and land management that aims to restore and enhance the health of ecosystems while also improving agricultural productivity and resilience (Sahu and Das 2020; Mishra et al. 2022). At its core, RA seeks to work with natural processes rather than against them, promoting soil health, biodiversity, water retention, and carbon sequestration. It seeks to create farming systems that are not only economically viable but also environmentally sustainable and socially responsible. By prioritizing soil health and ecosystem resilience, regenerative agriculture offers a promising pathway towards addressing the challenges of climate change, biodiversity loss, and food insecurity (Mishra et al. 2022).

The core principles of RA include minimizing soil disturbance, cover cropping, crop rotation, integrating diverse crops and livestock, conservation tillage, adding organic matter to the soil, integrating trees and shrubs, and emphasizing water management practices (Lal 2015a; Sahu and Das 2020). RA integrates principles of One Health, recognizing the interconnectedness of soil, plant, animal, and human health within agricultural systems. By adopting holistic and sustainable approaches to farming and land management, it supports ecosystem health and resilience while offering benefits for food security, nutrition, and environmental sustainability.

Despite data gaps, there is growing evidence that unsustainable agricultural practices harm soil health, crop quality, and ecosystem resilience. Sustainable solutions are needed to conserve soils and mitigate the wide-ranging detrimental impacts of conventional agriculture, including soil degradation, loss of biodiversity, and pollution. Transitioning to regenerative agriculture is essential to promote ecosystem health and resilience.

The aim of the study is to investigate the role of regenerative agriculture practices in promoting holistic health within agricultural systems and restoring soil health. Through an exploration of regenerative agricultural techniques such as minimal soil disturbance, cover cropping, crop rotation, and integration of diverse crops and livestock, the study aims to demonstrate how these practices contribute to the interconnectedness of soil, plant, animal, and human health within agricultural ecosystems. By assessing the impacts of regenerative agriculture on soil health parameters, biodiversity, water quality, and human well-being, the study seeks to provide evidence for the effectiveness of regenerative practices in mitigating the adverse effects of conventional agriculture, promoting ecosystem resilience, and fostering sustainable food systems (Fig. 1.1).

Methodology

Description of the Systematic Review Process for Selecting Empirical and Theoretical Studies

This review article focuses on investigating the role of RA as a channel for One Health principles and soil health restoration. Initially, research gaps and objectives were established to guide the review process, aiming to discern the evidence supporting the relationship between RA practices and their impact on soil health, ecosystem dynamics, and human-animal-plant interactions. A comprehensive search strategy was developed, incorporating keywords and controlled vocabulary terms related to regenerative agriculture, One Health, and soil restoration across various databases, including academic journals, conference proceedings, and grey literature sources. Inclusion and exclusion criteria were defined to screen and select studies based on relevance, publication type, study design, and focus on regenerative agriculture practices. Initial screening involved assessing titles and abstracts, followed by a full-text review of selected studies to determine eligibility. Data extraction was conducted systematically to collect relevant information from included studies, such as study characteristics and key findings. Synthesized findings were analyzed to identify patterns, themes, and relationships across studies, addressing the research

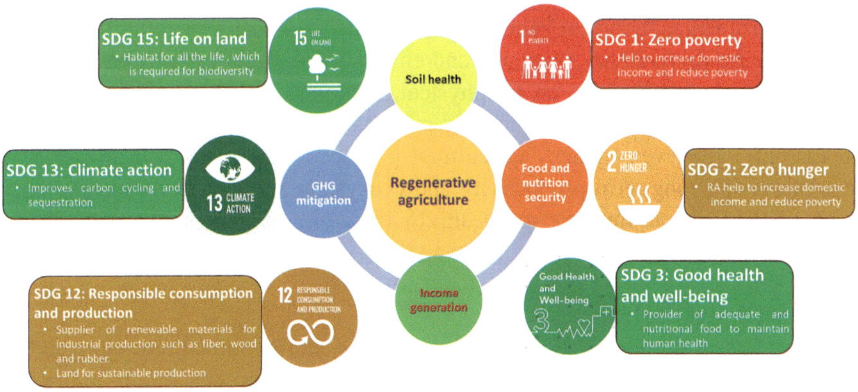

Fig. 1.1 An overview of regenerative agriculture and its correlation with various social development goals

questions and objectives of the review. The results were reported following established reporting guidelines, with a focus on transparency, reproducibility, and dissemination of findings to academic, policy, and practitioner audiences interested in regenerative agriculture, One Health, and sustainable land management.

Criteria for Inclusion

The criteria for inclusion in this study encompass peer-reviewed articles that specifically address comparative studies between regenerative and conventional farming methods, with a particular focus on soil health and One Health outcomes. These studies should rigorously analyze and compare the effects of regenerative agricultural practices and conventional farming techniques on soil health parameters, such as soil organic matter content, microbial diversity, nutrient availability, and soil erosion rates. Additionally, the articles should investigate the broader One Health implications, considering the interconnectedness of soil, plant, animal, and human health within agricultural systems. Studies exploring the effects of farming practices on ecosystem resilience, biodiversity, water quality, and human nutrition are particularly interesting. The selected articles should provide empirical evidence and robust data to support their findings, contributing to a comprehensive understanding of the benefits and challenges associated with regenerative agriculture compared to conventional farming methods.

Core Concepts of Regenerative Agriculture

Soil Health

Soil health is crucial for maintaining a functioning ecosystem and sustaining agricultural productivity. It refers to the soil's ability to support plant growth, maintain air and water quality, and provide essential services to ecosystems. The Intergovernmental Technical Panel on Soils defines soil health as the soil's capacity to sustain the productivity, diversity, and environmental services of terrestrial ecosystems. Soil health relies on its physical, chemical, and biological properties, including texture, pH, soil organic matter (SOM), microbial diversity, and nutrient availability. However, modern agricultural practices have led to soil degradation, compromising long-term productivity and ecosystem services. Regenerative agriculture (RA) aims to improve soil health by increasing organic matter and enhancing fertility and productivity.

Increased Soil Carbon

Soil serves as a significant reservoir of carbon (C), storing three times more carbon than the atmosphere. Soil organic carbon (SOC) plays a crucial role in improving soil structure, fertility, nutrient availability, and water retention. Increasing SOC is also considered a solution for mitigating climate change. The "4 per 1000" initiative suggests that increasing SOC by 0.4% annually in the top 30–40 cm of soil across all land uses could absorb a substantial amount of CO_2 emissions, benefiting soil health and food security (Bass et al. 2016; Pörtner et al. 2022). Although there are criticisms and uncertainties surrounding this initiative, increasing SOC even by a small percentage can have significant benefits for soil health and agricultural productivity.

Minimum/No Tillage

Minimum or no tillage is a key practice in RA that aims to minimize soil disturbance, promote fungal hyphae proliferation, and enhance nutrient cycling. This practice reduces carbon dioxide (CO_2) emissions and conserves water resources. Minimum tillage, combined with residue retention in a double-cropping system, is the most promising strategy for increasing SOC stocks in croplands. However, the effectiveness of conservation tillage practices in increasing SOC depends on various factors, including precipitation, soil depth, crop yield, and stubble retention (Yu et al. 2020; Singh et al. 2023).

Studies have shown that soil disturbance caused by tillage is the primary cause of SOC loss, and transitioning from conventional to conservation tillage practices can lead to substantial SOC sequestration (Song et al. 2019; McClelland et al. 2021). In some cases, long-term tillage practices produce more noticeable changes in SOC than short-term practices. Soil carbon sequestration under no-till farming enhances soil chemical fertility, promotes higher biological activity, and increases soil productivity compared to conventional tillage. However, while no-till farming benefits soil quality, its role in mitigating climate change may be exaggerated (Miura et al. 2013; Pandey et al. 2014; White et al. 2020).

Cover Crop

The second and third principles of regenerative agriculture involve keeping soil covered and having living roots in the soil throughout the year. One way to achieve this is by incorporating cover crops into the farming system. Cover crops are plants grown between main crops to cover and keep living plants on the soil during non-cash-cropping periods. They can be planted after harvest or under-seeded with perennial crops that continue to grow after the main crop is harvested. Cover crops can be single species or multispecies blends. While single cover crop species are easy to manage, a mix of species may provide all the benefits of each species in the mix (Lienhard et al. 2013; White et al. 2020).

Multispecies cover crops, including legumes, are believed to improve ecosystem functions such as biological nitrogen fixation, microbial diversity, compaction reduction, attraction of beneficial insects, weed suppression, regulation of soil temperature, and increased water infiltration. Additionally, cover crops aid in carbon sequestration, and widespread adoption could reduce agricultural greenhouse gas emissions by 10%, similar to using no-till or other cropping practices. One key benefit of cover crops is enhanced microbial biomass through the addition of extra organic matter in the soil. However, significant increases in soil carbon may take several years to occur (Jian et al. 2020; McClelland et al. 2021).

Various responses to soil organic carbon (SOC) accumulation have been reported globally in different agroclimatic regions. Cover cropping was shown to increase SOC storage in a temperate humid region of North America, improving soil functionality, but profitability was dependent on the production system used. The mechanism of SOC storage with cover crops is not fully understood and could be due to either below- or aboveground biomass input or rhizodeposition. Soil carbon accumulation with cover crops has been linked to soil texture, with an increase in soil carbon more likely to occur in clay soils. Studies in Argentina have shown that cover crops grown on fine- and coarse-textured soils accumulate more soil carbon. Cover crops can help eroded soils with low carbon content accumulate more carbon, with benefits more visible in no-till systems due to slower residue decomposition compared to conventional tillage (Kaye and Quemada 2017; Jordon et al. 2022).

Cover cropping is widely recommended to prevent soil erosion and loss of soil carbon. Cover crop roots anchor and stabilize soil, create soil aggregates, and increase carbon inputs from root and shoot cover crop residues. These changes can lead to increased uptake of available nitrogen and decreased nitrous oxide and methane emissions. Cover cropping is also suggested as an important practice for adapting to future climate risks due to its ability to decrease erosion and improve soil fertility and health (Lal 2015a, b; Kaye and Quemada 2017).

Crop Diversification/Rotation

Crop rotation involves planting a planned sequence of crops over time in the same field. This practice has been used for centuries and is still important in modern farming because it offers several benefits. Crop rotation helps reduce crop diseases, weeds, and pests while increasing crop yields (Bullock 1992; Shah et al. 2021). It is an ancient method that diversifies crops, providing nutritional benefits and breaking the cycle of pests, diseases, and weeds, which ultimately improves yield and profit. However, during the mid-twentieth century, crop rotation was largely replaced by monoculture farming due to heavy use of fertilizers and pesticides, improved crop varieties, and economic factors. This led to land degradation and loss of soil organic matter (SOM) (Bullock 1992). Nowadays, crop rotation is gaining recognition for its potential to enhance soil quality, crop yield, and water-use efficiency (Shah et al. 2021). Deep-rooted perennials, like those used in crop rotation, can help sequester soil organic carbon (SOC) at greater depths, especially in less productive soils. Factors such as soil depth and rainfall influence SOC levels, with annual pastures having higher SOC than mixed or continuous cropping systems. Recent studies suggest that crop rotation may increase SOC sequestration rates by improving soil aggregation and increasing plant carbon inputs (King and Blesh 2018; Shah et al. 2021). Furthermore, crop rotation enhances agricultural climate resilience by diversifying agricultural systems (Yu et al. 2022). Agroforestry is a land-use practice that involves intentionally incorporating trees, shrubs, palms, bamboo, and other woody perennials into croplands. It offers numerous benefits and ecosystem services, such as improved water and air quality, soil enrichment, and biodiversity conservation (Jose 2009). Research indicates that agroforestry can enhance soil carbon sequestration by improving nutrient and water intake efficiencies, reducing nutrient leaching and soil erosion, and enhancing soil physical and biological properties (Abbas et al. 2017; Cardoso et al. 2017; Muchane et al. 2020). While some studies have examined the effects of agroforestry on plant shoot and root biomass carbon (Zomer et al. 2016), limited research focuses on its impact on soil carbon.

Water Conservation

Regenerative agriculture focuses on actively restoring soil health and ecosystem function while promoting sustainable farming practices. Within this framework, water conservation emerges as a crucial component, contributing to soil health, biodiversity, and overall ecosystem resilience. Two primary techniques for water conservation in regenerative agriculture are rainwater harvesting and drip irrigation (Sanchis et al. 2012; Pandey et al. 2014).

Rainwater harvesting involves the collection and storage of rainwater to supplement irrigation needs and enhance soil moisture content. This technique aligns perfectly with regenerative agriculture principles by leveraging natural processes to improve water efficiency on farms. Farmers can capture rainwater from rooftops, surfaces, or designated catchment areas, reducing dependence on groundwater or surface water sources. Stored rainwater can then be used during dry periods to irrigate crops, replenish soil moisture, and support ecosystem health (Arunrat et al. 2022).

Drip irrigation is another essential practice in regenerative agriculture for water conservation. Unlike traditional irrigation methods that often result in water wastage through runoff and evaporation, drip irrigation delivers water directly to plant roots in a controlled and efficient manner. This minimizes water loss, promoting soil health and biodiversity while reducing the risk of soil erosion and nutrient leaching. In regenerative agriculture systems, drip irrigation can be integrated with soil-building practices such as cover cropping and mulching to further enhance water retention and plant resilience (Janz et al. 2019).

Both rainwater harvesting and drip irrigation contribute significantly to the overall sustainability of regenerative agriculture systems. By optimizing water use, these techniques support the regeneration of soil health and ecosystems, leading to increased productivity and resilience in the face of climate variability. Additionally, they help mitigate the impact of droughts and water scarcity by providing farmers with reliable water sources that are less susceptible to depletion. These practices demonstrate the proactive approach of regenerative agriculture towards sustainable water management and ecosystem conservation.

Enhancing Biodiversity

Soil biodiversity, which encompasses a diverse range of organisms, is fundamental for the functioning and health of terrestrial ecosystems. In the intricate soil matrix, various organisms exist, from microscopic bacteria and fungi to larger invertebrates like earthworms and nematodes. Bacteria and archaea dominate the microbial community, playing crucial roles in nutrient cycling, organic matter decomposition, and soil structure formation. Fungi, including both saprophytic and mycorrhizal species, aid in organic matter decomposition and facilitate nutrient uptake by plants. Protists, nematodes, and microarthropods regulate microbial populations and nutrient dynamics through predation and grazing. Larger organisms such as earthworms and macroarthropods improve soil aeration, nutrient mixing, and organic matter decomposition through their burrowing and feeding activities. Each group of organisms interacts within the soil ecosystem, forming intricate food webs that contribute to ecosystem resilience and stability. Overall, the diversity and composition of soil organisms are crucial for maintaining soil fertility, productivity, and provision of ecosystem services essential for sustaining life on Earth. Soil biodiversity is crucial for organic matter breakdown, nutrient cycling, and soil structure, which are essential for soil fertility and ecosystem health. Microorganisms play a significant role in biochemical transformations, with 80–90% of total soil metabolism attributed to them. Agricultural intensification is a major cause of soil biodiversity loss, but conservation practices can help mitigate these effects and promote sustainable farming and ecosystem resilience (Schreefel et al. 2020; Wurz et al. 2022).

Traditional agricultural practices heavily rely on inorganic fertilizers, pesticides, and growth regulators to enhance crop yields. However, the extensive use of these agrochemicals is increasingly recognized for its detrimental impact on the environment. Chemical fertilizers became widespread following the green revolution in the early 1960s due to their perceived ability to boost crop productivity. However, excessive use of agrochemicals leads to a significant decline in crop yields, pollutes soil ecosystems, and contaminates water bodies (Hok et al. 2021).

Tillage, chemical fertilizers, pesticides, and residue burning profoundly impact soil biodiversity. Tillage disrupts soil structure and organic matter, leading to habitat loss and decreased microbial diversity (Lienhard et al. 2013). Chemical fertilizers and pesticides can harm soil microbial communities by altering soil pH, degrading organic matter, and reducing microbial abundance and diversity. Pesticide residues can persist in the soil, further impacting soil biota and ecosystem functioning. Residue burning depletes organic carbon and essential nutrients for soil organisms, while also releasing harmful pollutants into the atmosphere. These practices collectively contribute to the degradation of soil biodiversity, compromising soil health, ecosystem resilience, and agricultural sustainability. Implementing conservation practices such as reduced tillage, organic fertilizers, integrated pest management, and residue retention can mitigate these negative impacts and promote soil biodiversity conservation.

The burning or removal of crop residue directly affects the soil health. Crop residue retention and/or incorporation improve or maintain soil quality and biodiversity. These conservation agricultural practices promote soil fertility by increasing soil carbon, nitrogen, and microbial biomass, benefiting agroecosystems in the long term (Fig. 1.2).

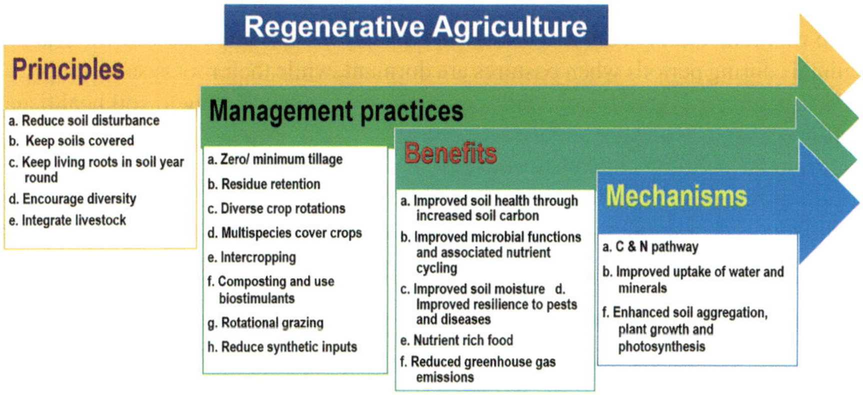

Fig. 1.2 Regenerative agriculture principles, practices, and purported benefits and mechanisms to improve soil health

Integrated Crop and Livestock Management

Integrated crop and livestock management (ICLM), a cornerstone of regenerative agriculture, represents a holistic approach to farming that combines crop production with animal grazing to achieve multiple benefits. This integrated system fosters synergies between crop and livestock activities, emphasizing nutrient cycling and pest control to enhance overall farm sustainability. One key aspect of ICLM is its focus on nutrient cycling, where grazing animals play a vital role. As animals graze, they recycle organic matter through their manure, which serves as a natural fertilizer for crops. This organic fertilizer enriches the soil with essential nutrients such as nitrogen, phosphorus, and potassium, promoting healthy plant growth and improving soil fertility. Moreover, the incorporation of animal manure into the soil enhances soil structure and water-holding capacity, leading to increased crop yields and reduced dependency on synthetic fertilizers (Giller et al. 2021). In addition to nutrient cycling, ICLM practices leverage the grazing behavior of animals for effective pest control. Grazing animals, such as cattle or sheep, can help manage weeds and pests by consuming unwanted vegetation. By grazing on weeds, they reduce competition for resources and help maintain a balanced ecosystem within the farm. Furthermore, the disturbance caused by grazing animals to the soil surface can disrupt pest habitats and reduce pest populations naturally, thereby minimizing the need for chemical pesticides.

The integration of crop production and grazing activities in ICLM systems creates several synergies. Rotational grazing systems, for example, optimize pastureland utilization by moving animals between different areas, minimizing overgrazing, and allowing forage to regrow. This approach not only supports healthy pasture ecosystems but also provides farmers with a consistent source of high-quality forage for their animals. Moreover, integrating cover crops and forage crops into crop rotations offers additional benefits. These crops can serve as feed for grazing animals during periods when pastures are dormant, while their root systems improve soil structure, prevent erosion, and sequester carbon, contributing to soil health and fertility.

From an environmental perspective, ICLM practices align closely with the principles of regenerative agriculture. By reducing reliance on synthetic inputs such as fertilizers and pesticides, integrated systems minimize environmental impacts and promote sustainable land management. Enhanced nutrient cycling, soil conservation, and biodiversity conservation contribute to improved ecosystem resilience and long-term farm viability (Alexanderson et al. 2023) (Table 1.1).

Table 1.1 Agronomic principles and practices considered to be part of RA and their potential impacts on restoration of soil health and reversal of biodiversity loss

Principles	Practices	Restoration of soil health	Reversal of soil biodiversity
Reduce soil disturbances	Zero/minimum tillage, residue retention	***	**
Permanent soil cover	Mulch, cover crops, permaculture	***	**
Increase soil C and sequester carbon	Biochar, compost, green manures, animal manures, agroforestry, tree crops	***	**
Depending more on organic manure	Animal manures, compost, bio-fertilizers, green manures, and crop residues maintain living roots in soil, reduce dependence on mineral fertilizers, etc.	***	***
Foster plant diversity	Diverse crop rotations, multispecies cover crops, agroforestry	**	***
Integrate livestock	Rotational grazing, pasture cropping, silvo-pasture	**	**
Avoid chemical fertilizers, herbicides, and pesticides	Diverse crop rotations, multispecies cover crops, integrated pest management	**	***
Adoption of water conservation techniques	Different water-shaving strategies such as rainwater harvesting, drip irrigation, and soil ameliorants (biochar, compost, manures, etc.)	***	**

Based on Giller et al. (2021)

The One Health Framework

Detailed Exploration of How Healthier Soils Contribute to Healthier Crops, Which Support Healthier Animal and Human Populations

In regenerative agriculture, the focus on healthier soils as a foundation for sustainable food production is paramount. Healthy soils contribute to healthier crops, which in turn support healthier animal and human populations. This intricate relationship is at the core of regenerative agricultural practices, which aim to enhance soil health through natural processes and sustainable management techniques.

Improved Soil Structure and Nutrient Availability

Regenerative agriculture plays a crucial role in maintaining healthy soils. Practices like minimal tillage, cover cropping, and crop rotation help improve soil structure and reduce compaction, allowing roots to grow well and absorb nutrients efficiently. These practices also enhance organic matter decomposition and nutrient cycling, ensuring that essential nutrients like nitrogen, phosphorus, and potassium are available for plant growth, which is vital for high productivity (Kibblewhite et al. 2008; Lal 2015b).

Enhanced Soil Fertility and Microbial Activity

Healthy soils contain diverse microbial communities that are essential for nutrient cycling, disease prevention, and maintaining soil structure. Regenerative agriculture methods, such as cover cropping, crop rotation, and adding organic materials, promote the diversity and activity of these microbes. This encourages beneficial relationships between plants and soil microbes, leading to better nutrient availability and soil fertility. Additionally, healthy soils have more organic matter, which stores nutrients and provides energy for soil organisms. Regenerative agriculture plays a crucial role in enhancing these soil benefits (Tahat et al. 2020; Meena et al. 2024).

Increased Crop Resilience to Environmental Stressors

Healthy soils play a vital role in crop resilience by improving water infiltration, retention, and drainage, which helps to reduce the impact of drought and waterlogging. Regenerative practices such as mulching, cover cropping, and agroforestry enhance soil water-holding capacity and reduce soil erosion, thereby lessening the effects of extreme weather events (Jose 2009; Yu et al. 2022). Moreover, healthy soils support diverse and robust plant communities that can better withstand pest and disease pressure. Crops grown in healthy soils tend to be more nutrient-denser and flavorful compared to those grown in degraded soils. The availability of essential nutrients in healthy soils leads to higher nutrient levels in crops, enhancing their nutritional value. Additionally, a balanced nutrient supply and optimal plant health improve crop quality, taste, and aroma, which contributes to consumer satisfaction and overall well-being. Healthier crops produced in regenerative agricultural systems provide nutritious feed for livestock, supporting their health and productivity. Livestock raised on nutrient-rich forage and pasturelands often show better growth rates, reproductive performance, and disease resistance. Consequently, consuming nutrient-dense crops and animal products from regenerative systems contributes to the overall health and well-being of human populations (Brown et al. 2021). Diets rich in fruits, vegetables, grains, and pasture-raised meats from regenerative systems provide essential vitamins, minerals, and antioxidants, reducing the risk of chronic diseases and promoting optimal health.

Ecological Health as a Critical Component of Sustainable Agriculture

Ecological health is a critical component of sustainable agriculture, and regenerative agriculture plays a significant role in promoting and maintaining it. One of the key aspects of ecological health in agriculture is the prevention of soil erosion and degradation, which can have profound impacts on the environment, agricultural productivity, and food security. Regenerative agriculture offers innovative practices aimed at restoring and preserving soil health, thus mitigating erosion and degradation (Brown et al. 2022; Mishra et al. 2022; Gordon et al. 2023).

Firstly, regenerative agriculture prioritizes soil health as the foundation of sustainable farming systems. By implementing practices such as minimal tillage, cover cropping, crop rotation, and use of organic amendments, regenerative farmers enhance soil structure, organic matter content, and microbial diversity. These practices increase soil stability and water infiltration rates, reducing the risk of erosion caused by wind and water runoff (Tahat et al. 2020).

Moreover, regenerative agricultural practices promote the development of robust root systems, which further stabilize the soil and prevent erosion. Diverse crop rotations and cover cropping strategies help maintain continuous ground cover, reducing the exposure of bare soil to erosive forces. Additionally, the incorporation of perennial crops and agroforestry systems in regenerative farming landscapes provides long-term soil protection and stabilization (O'Donoghue et al. 2024).

Furthermore, regenerative agriculture emphasizes the importance of holistic land management approaches that consider the interactions between soil, plants, animals, and the surrounding ecosystem. By integrating livestock into cropping systems and implementing rotational grazing practices, regenerative farmers improve soil fertility and structure through the natural recycling of nutrients and the enhancement of soil microbial activity. This not only helps prevent erosion but also contributes to the restoration of degraded soils (Jayasinghe et al. 2023).

In terms of One Health outcomes, regenerative agriculture benefits not only soil health but also plant, animal, and human health. Healthy soils support thriving plant growth, leading to increased crop yields and nutrient-dense food production. By reducing the need for synthetic fertilizers and pesticides, regenerative practices minimize environmental pollution and mitigate the risks of human exposure to harmful chemicals (Khangura et al. 2023; Buckton et al. 2023).

Comparative Studies

The results from various studies examining the effects of regenerative agricultural practices on soil health are presented below (Table 1.2). In a study by Dey et al. (2021), conducted over 12 years on clay loam soil, the adoption of regenerative practices in a rice-wheat cropping system led to a 20.1% increase in SOC compared to conventional practices. Similarly, Jat et al. (2019a, b) observed significant increases in SOC (66.7% and 67.3%, respectively) in sandy loam soils over 4 and 6 years of implementation of regenerative practices in rice-wheat systems. In another study by Dey et al. (2016) on sandy clay loam soil, a 4.3% increase in SOC was recorded over 3 years. Bera et al. (2017) found a 15.5% increase in SOC in sandy loam soil over 5 years. Dhaliwal et al. (2020) reported a 29.1% increase in SOC in sandy loam soil over 5 years. Sharma et al. (2021) observed an 8.11% increase in SOC in sandy loam soil over 5 years. Jat et al. (2018) recorded a 71.1% increase in SOC in sandy loam soil over 4 years. Parihar et al. (2019, 2020) reported increases of 22.1% and 18.4%, respectively, in sandy loam soil over 5 and 4 years. Singh et al. (2016) observed a 37.7% increase in SOC in sandy loam soil over 5 years in a rice-maize cropping system. Alam et al. (2018) found a 52.7% increase in SOC in silt loam soil over 5 years in a mustard-rice cropping system. Kumar and Nath (2019) reported a 39.7% increase in SOC in sandy loam soil over 7 years in a rice-based cropping system. Nath et al. (2017) recorded a 7.9% increase in SOC in sandy loam soil over 2 years in a wheat-mung bean cropping system. Somasundaram et al. (2019) observed a 3.8% increase in SOC in Vertisol over 3 years in a soybean-wheat cropping system. Parihar et al. (2016, 2018) reported increases of 34.8% and 37.2%, respectively, in sandy loam soil over 7 and 6 years in maize-based cropping systems. Yadav et al. (2019) observed a 2.8% increase in SOC in sandy clay loam soil over 3 years in a rice-rice cropping system. Dash et al. (2017) found a 9.7% increase in SOC in sandy clay loam soil over 5 years in a rice-rice cropping system. These findings underscore the positive impact of regenerative agricultural practices on SOC accumulation across various cropping systems and soil types, supporting the importance of such practices for enhancing soil health and sustainability.

The majority of studies conducted over periods ranging from 3 to 10 years have demonstrated positive effects of RA on soil health parameters. The soil health effects of RA vary significantly between sites, depending on local conditions such as soil type, production system, or climate. However, most of the studies suggest that the RA system should be promoted to achieve soil health benefits and higher crop productivity with lower input costs.

Benefits of a Regenerative Approach

A regenerative approach to agriculture offers a multitude of benefits, particularly in terms of ecological resilience. By focusing on practices that enhance ecosystem services and biodiversity, regenerative agriculture contributes to the sustainability and resilience of natural systems. Here is an elaborate exploration of the benefits:

Table 1.2 Effect of regenerative agriculture on soil organic carbon (SOC) content in different cropping systems

Studies	Cropping system	Soil type	Duration (years)	Soil depth (cm)	SOC (g/kg) Conventional practice	Regenerative practice	Δ (%)
Dey et al. (2021)	Rice-wheat	Clay loam	12	0–15	22.9	27.5	20.1
Jat et al. (2019a)	Rice-wheat	Sandy loam	4	0–15	4.5	7.5	66.7
Jat et al. (2019b)	Rice-wheat	Sandy loam	6	0–15	4.9	8.2	67.3
Dey et al. (2016)	Rice-wheat	Sandy clay loam	3	0–15	6.95	7.25	4.3
Bera et al. (2017)	Rice-wheat	Sandy loam	5	0–7.5	5.94	6.86	15.5
Dhaliwal et al. (2020)	Rice-wheat	Sandy loam	5	0–15	4.8	6.2	29.1
Sharma et al. (2021)	Rice-wheat	Sandy loam	5	0–7.5	14.8	16	8.11
Jat et al. (2018)	Rice-wheat	Sandy loam	4	0–15	4.5	7.7	71.1
Parihar et al. (2019)	Rice-wheat	Sandy loam	5	0–7.5	5.53	6.75	22.1
Parihar et al. (2020)	Rice-wheat	Sandy loam	4	0–15	5.11	6.05	18.4
Singh et al. (2016)	Rice-maize	Sandy loam	5	0–15	4.3	5.92	37.7
Alam et al. (2018)	Mustard-rice	Silt loam	5	0–10	6.56	10.02	52.7
Kumar and Nath (2019)	Rice based	Sandy loam	7	0–20	3.73	5.21	39.7
Nath et al. (2017)	Wheat-mung bean	Sandy loam	2	0–15	15.2	16.4	7.9
Somasundaram et al. (2019)	Soybean-wheat	Vertisol	3	0–15	5.3	5.5	3.8
Parihar et al. (2016)	Maize based	Sandy loam	7	0–15	4.8	6.5	34.8
Parihar et al. (2018)	Maize based	Sandy loam	6	0–5	5.46	7.68	37.2
Yadav et al. (2019)	Rice-rice	Sandy clay loam	3	0–20	10.6	10.9	2.8
Dash et al. (2017)	Rice-rice	Sandy clay loam	5	0–30	16.9	15.4	9.7

Soil Health

Regenerative practices such as minimal soil disturbance, cover cropping, and composting improve soil structure, fertility, and nutrient cycling. Healthy soils support diverse microbial communities, which in turn promote nutrient availability and plant health. This resilience in soil health enables agricultural systems to withstand disturbances such as droughts, floods, and erosion, thereby enhancing ecological resilience. Increased carbon sequestration: One of the key features of regenerative agriculture is its ability to sequester carbon from the atmosphere into the soil. Practices like rotational grazing, agroforestry, and conservation tillage promote the accumulation of organic matter in the soil, leading to carbon storage (Jat et al. 2018; Mishra et al. 2024). This not only mitigates climate change by reducing atmospheric carbon dioxide levels but also improves soil fertility and water retention. Biodiversity conservation: Regenerative farming methods create diverse and resilient ecosystems that support a wide range of plant and animal species. By incorporating diverse crop rotations, agroforestry, and hedgerows, regenerative farmers provide habitats and food sources for pollinators, birds, and beneficial insects. This biodiversity enhances ecosystem resilience by increasing stability and reducing the likelihood of pest outbreaks and crop failures (Schreefel et al. 2020).

Regenerative practices such as contour plowing, mulching, and cover cropping help to conserve water, improve water infiltration, and reduce soil erosion. Healthy soils with high organic matter content can hold more water, making agricultural systems more resilient to droughts and floods. Additionally, regenerative agriculture reduces the need for irrigation, thereby conserving water resources and minimizing environmental impact (Schreefel et al. 2020; Vamshi et al. 2024).

Regenerative agriculture minimizes the use of synthetic fertilizers, pesticides, and herbicides, which can have detrimental effects on soil health, water quality, and biodiversity. Instead, farmers employ natural and organic methods such as integrated pest management, crop rotation, and biological controls to manage pests and diseases. This reduces chemical runoff into waterways and minimizes the risk of environmental pollution (Sahu and Das 2020; Rehberger et al. 2023).

While transitioning to regenerative practices may require initial investments and changes in management techniques, it often leads to improved long-term economic viability. Healthy soils result in higher crop yields, reduced input costs, and decreased reliance on external inputs. Moreover, diversified farming systems with multiple revenue streams, such as agroforestry and value-added products, can provide additional income opportunities for farmers (Lal 2006; Song et al. 2019; Wurz et al. 2022).

Regenerative agriculture helps mitigate the impacts of climate change by sequestering carbon, reducing greenhouse gas emissions, and enhancing resilience to extreme weather events. By building soil organic matter and increasing biodiversity, regenerative practices create agricultural systems that are better able to adapt to changing environmental conditions, such as heat waves, droughts, and heavy rainfall (Lal 2015b; Kayc and Quemada 2017).

Economic Viability

A regenerative approach to agriculture offers numerous benefits, including economic viability, which encompasses cost-effectiveness, profitability, and long-term financial advantages for farmers. Firstly, regenerative practices can lead to cost savings by reducing the reliance on external inputs such as synthetic fertilizers, pesticides, and herbicides. By enhancing soil health through techniques like cover cropping, crop rotation, and minimal tillage, regenerative farmers can improve nutrient cycling and reduce the need for expensive chemical inputs. This not only lowers production costs but also mitigates the financial risks associated with input price volatility (Newton et al. 2020; Rehberger et al. 2023).

Moreover, regenerative agriculture promotes higher yields and increased resilience to environmental stresses. Healthy soils support vigorous plant growth, leading to higher crop productivity over time. By diversifying crop rotations and incorporating cover crops, farmers can better manage risks associated with pests, diseases, and extreme weather events, ultimately safeguarding their income streams. In addition to immediate cost savings and increased productivity, regenerative practices offer long-term financial benefits for farmers. One key advantage is improved soil health and fertility (Schreefel et al. 2020; Jian et al. 2020; McCauley and Barlow 2023). Healthy soils are more resilient to erosion, retain water more effectively, and provide a stable foundation for sustainable agricultural production over the long term. As a result, farmers practicing regenerative agriculture can maintain or even enhance the productivity of their land without the need for expensive inputs or costly interventions (Kremen and Miles 2012).

Furthermore, regenerative farming systems can create additional revenue streams for farmers. For example, integrating livestock into crop rotations can provide opportunities for diversification and value-added products such as grass-fed meat and dairy. Agroforestry practices, such as planting trees on marginal lands, can generate income from timber, fruit, or non-timber forest products while also enhancing biodiversity and ecosystem services (Janz et al. 2019; Yu et al. 2020).

The economic viability of regenerative agriculture is further bolstered by market demand for sustainably produced food and fiber. Consumers increasingly value products that are grown using environmentally friendly practices, and many are willing to pay a premium for organic, regeneratively grown, or locally sourced goods. By adopting regenerative practices, farmers can tap into these market opportunities, differentiate their products, and capture higher prices, thus improving their profitability (Rehberger et al. 2023; O'Donoghue et al. 2024).

Social Equity and Public Health

A regenerative approach to agriculture not only improves environmental sustainability but also fosters social equity and public health by reducing chemical exposure, promoting local food production, empowering communities, creating economic opportunities, and preserving cultural heritage. By prioritizing the well-being of people and the planet, regenerative agriculture offers a holistic solution to the challenges facing modern food systems (Buckton et al. 2023).

Regenerative farming practices prioritize natural methods of pest control and soil fertility enhancement, reducing the reliance on synthetic pesticides, herbicides, and fertilizers. By minimizing chemical inputs, these practices lower the exposure of farmworkers, nearby communities, and consumers to potentially harmful substances. This reduction in chemical exposure contributes to improved public health outcomes by lowering the risk of pesticide-related illnesses and chronic diseases associated with pesticide residues in food (Evans et al. 2015; Hok et al. 2021).

Regenerative agriculture emphasizes local food systems, where food is produced closer to where it is consumed. By supporting local farmers and food producers, regenerative practices contribute to food sovereignty and security, ensuring that communities have access to fresh, nutritious, and culturally appropriate food. Increased local food production also reduces the carbon footprint associated with transportation and distribution, leading to lower greenhouse gas emissions and mitigating climate change (Sahu and Das 2020; Arunrat et al. 2022).

Adopting regenerative practices fosters community engagement and empowerment by reconnecting people with the land and food production processes. Community-supported agriculture (CSA) models, farmers' markets, and farm-to-school programs are examples of initiatives that promote direct interaction between producers and consumers, building trust and strengthening local food systems. By actively participating in food production and distribution, communities gain a sense of ownership and pride in supporting sustainable agriculture practices (Brown et al. 2021; Bhaduri et al. 2022; Mishra et al. 2022).

Regenerative agriculture creates opportunities for job creation and economic development, particularly in rural and marginalized communities. By diversifying farming practices and incorporating agroecological principles, regenerative farms can support a wider range of livelihoods, including small-scale farming, agroforestry, and value-added food processing. Additionally, the shift towards local food systems stimulates local economies by keeping more money circulating within the community and reducing dependence on external inputs and corporate supply chains (Mishra et al. 2022; Beacham et al. 2023).

Many regenerative farming practices are rooted in traditional knowledge and indigenous agricultural practices that have sustained communities for generations. By preserving and revitalizing these cultural heritage practices, regenerative agriculture supports cultural identity and resilience, particularly among indigenous and minority communities. Recognizing and valuing diverse farming traditions also promote social equity by acknowledging the contributions of historically marginalized groups to agriculture and food systems.

Barriers to Implementation

Financial and Structural Challenges

The implementation of regenerative agricultural practices faces several barriers, with financial and structural challenges being among the most prominent.

Financial Barriers

Transitioning to regenerative agriculture often requires significant up-front investments in infrastructure, equipment, and training. Farmers may need to purchase new machinery, such as no-till planters or cover crop seeders, which can be expensive. Additionally, implementing practices like rotational grazing may require changes to fencing and water systems. These initial costs can be prohibitive for many farmers, especially those operating on tight budgets (Vinke et al. 2017; Wurz et al. 2022).

Lack of Access to Capital

Access to capital for investment in regenerative practices can be limited, particularly for small-scale farmers or those without established credit. Traditional lending institutions may be hesitant to finance regenerative projects due to perceived risks or lack of familiarity with these approaches. Without access to affordable financing options, farmers may struggle to make the necessary investments in transitioning to regenerative practices (Grelet et al. 2021).

Market Uncertainty

Another financial barrier is the uncertainty surrounding market demand and pricing for regeneratively produced goods. While there is growing consumer interest in sustainable and ethically produced food, farmers may face challenges in accessing premium markets or securing stable prices for their products. Without assurance of a return on investment, farmers may be reluctant to adopt regenerative practices (Mishra et al. 2022; Beacham et al. 2023).

Structural Challenges

Many regions lack the necessary infrastructure to support regenerative agriculture. This includes access to processing facilities for value-added products, storage facilities for specialty crops, and transportation networks to bring products to the market. Without adequate infrastructure, farmers may struggle to realize the full potential of regenerative practices or face additional costs in accessing off-farm services (Brown et al. 2022; Beacham et al. 2023).

Limited Knowledge and Training

Farmers may lack the knowledge and training needed to successfully implement regenerative practices. Traditional agricultural education often emphasizes conventional farming methods, leaving farmers unprepared to adopt alternative approaches. Training programs and extension services focused on regenerative agriculture are limited in many areas, making it difficult for farmers to learn new techniques and best practices (Zazo-Moratalla et al. 2019).

Policy and Regulatory Hurdles

Existing policies and regulations may pose barriers to the adoption of regenerative practices. Subsidies and support programs often favor conventional farming methods, providing few incentives for farmers to transition to regenerative systems. Regulatory frameworks may also be outdated or incompatible with regenerative approaches, leading to conflicts or uncertainty for farmers seeking to implement new practices (Chabbi et al. 2017; Al-Kaisi and Lal 2020).

Policy Constraints

Regenerative agriculture, while offering numerous environmental, social, and economic benefits, faces significant barriers to widespread adoption, including policy constraints. Existing agricultural policies often fail to adequately support or incentivize regenerative practices, hindering their adoption and implementation (Richelle et al. 2018; Al-Kaisi and Lal 2020). One primary constraint is the historical focus of policies on conventional farming methods, prioritizing yield maximization and cost reduction without sufficient consideration for sustainability. Subsidies and support programs tend to favor monoculture cropping and chemical inputs over regenerative practices like crop rotation and agroforestry (Shin et al. 2020). Additionally, the lack of specific policies tailored to support regenerative agriculture discourages farmers, as they may perceive a lack of recognition and financial assistance for their efforts. Regulatory frameworks may also pose obstacles, with regulations related to land use, water rights, and environmental protection often not accommodating regenerative practices (Chabbi et al. 2017; de Bruyn et al. 2017). Moreover, the complexity and fragmentation of agricultural policy landscapes create confusion for farmers, while the lack of long-term policy stability deters investment in transitioning to regenerative methods. To address these challenges, policymakers should revise existing policies to better align with regenerative agriculture principles, provide targeted financial incentives, streamline regulations, and foster collaboration among stakeholders to encourage the transition to more sustainable practices (Pauli et al. 2016; Park et al. 2022) (Fig. 1.3).

Strategic Recommendations

Policy Reforms

To foster the adoption of regenerative agricultural practices and promote sustainable farming systems, policymakers should implement several policy reforms and supportive measures. These initiatives aim to incentivize farmers, improve access to

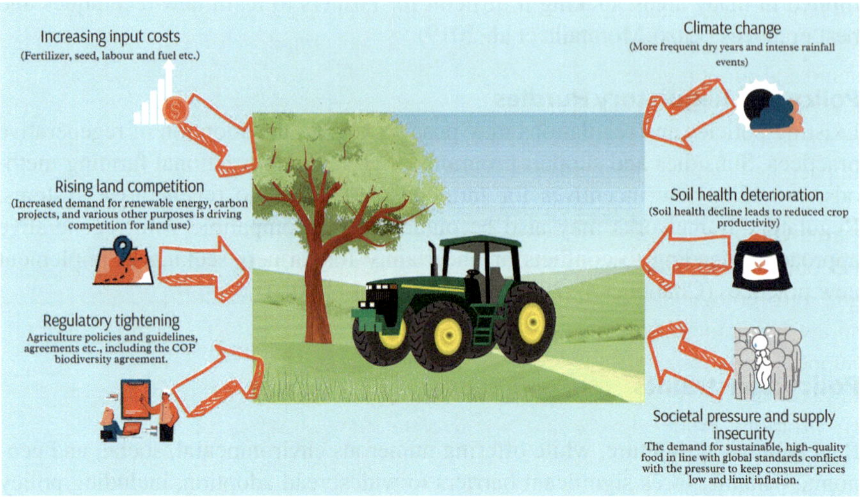

Fig. 1.3 Different barriers to implementation and adaptation of regenerative agriculture

resources, and create an enabling environment for regenerative agriculture. The following strategic recommendations are proposed:

Financial Incentives and Subsidies

To promote the adoption of regenerative practices among farmers, implementing financial incentives is crucial. Granting subsidies, tax credits, or grants can assist farmers in covering initial costs associated with transitioning to regenerative agriculture, thereby encouraging widespread adoption. Offering premium prices or marketing incentives for products produced using regenerative methods provides economic rewards for sustainable farming practices, further motivating farmers to embrace these approaches. In addition to financial incentives, allocating funding specifically for research and development in regenerative agriculture is essential. This funding should focus on various aspects, including soil health improvement, biodiversity conservation, and development of carbon sequestration technologies. By investing in research, we can advance our understanding of regenerative practices and their impacts, leading to more effective strategies for sustainable agricultural development.

Supportive Policies

Developing and enforcing policies that promote soil health and conservation are crucial for sustainable agriculture. These policies should encompass regulations on soil erosion, nutrient management, and pesticide use to foster a conducive environment for regenerative practices. Integrating regenerative agriculture principles into agricultural policy frameworks ensures that sustainability and environmental stewardship become central objectives. To support this integration, governments and agricultural authorities should provide technical assistance and extension services to farmers interested in transitioning to regenerative practices. This assistance should offer guidance on implementation and management, helping farmers understand and adopt regenerative techniques effectively. By implementing these policies and providing support services, governments can encourage widespread adoption of regenerative agriculture, leading to improved soil health, reduced environmental impact, and enhanced agricultural sustainability.

Land-Use Planning

Encourage sustainable land-use planning by prioritizing conservation, restoration, and agroecological approaches, including protecting agricultural land from urban sprawl, promoting agroforestry, and incentivizing land conservation practices. Support initiatives to ensure land tenure security for farmers, enabling long-term investments in regenerative practices without the risk of land loss or displacement. By promoting these measures, resilient agricultural systems can be fostered, supporting biodiversity, soil health, and ecosystem services while enhancing food security and community livelihoods.

Research and Education

Investing in research and education programs focusing on regenerative agriculture, including soil biology, agroecology, and ecosystem services, is crucial for generating scientific knowledge and building capacity among farmers and agricultural professionals. These initiatives should aim to deepen the understanding of regenerative principles and practices, exploring interactions between soil biology, plant health, animal welfare, and environmental sustainability. By conducting rigorous scientific studies, we can identify effective regenerative practices for different agroecological contexts. Additionally, education programs should disseminate knowledge and best practices through workshops, training sessions, and extension services to empower farmers to implement sustainable agricultural systems. Establishing demonstration farms and field trials is essential for showcasing the practical benefits of regenerative agriculture, serving as living laboratories where farmers can learn from real-world examples and gain confidence in adopting regenerative techniques such as soil health improvement, water conservation, biodiversity enhancement, and yield stability.

Market Access and Certification

To facilitate market access for regenerative agricultural products, it is crucial to establish certification standards and labels that recognize sustainable farming practices, encompassing key aspects such as soil health, biodiversity conservation, water management, and carbon sequestration. By certifying products meeting these criteria, consumers can easily identify and support regenerative farming practices. Additionally, promoting consumer awareness and demand is vital, achieved through education campaigns and marketing initiatives highlighting the environmental and social benefits of regenerative agriculture, such as improved soil health, biodiversity, reduced greenhouse gas emissions, and support for local communities. Marketing can emphasize regenerative products' unique qualities like superior taste, nutritional value, and ethical production methods, helping companies differentiate their products and attract environmentally conscious consumers. Overall, establishing certification standards and labels, along with education and marketing efforts, can drive market access and consumer demand for sustainable farming, benefiting farmers and contributing to broader environmental and social sustainability goals.

International Collaboration

To advance regenerative agriculture globally, active international collaboration and knowledge sharing are essential. Partnerships among governments, research institutions, NGOs, and agricultural stakeholders facilitate the exchange of best practices, innovative technologies, and effective policy approaches. This collaboration aims to share successful regenerative farming techniques, including experiences of farmers, insights from research, and field trials, to identify strategies for different contexts and scale up adoption. Additionally, international collaboration involves advocating for policies that support regenerative agriculture at global forums such as the United Nations Framework Convention on Climate Change (UNFCCC) and the Food and Agriculture Organization (FAO), promoting its integration into national and international policies related to climate change, biodiversity conservation, and food security. Advocacy efforts include ensuring that regenerative practices are included in Nationally Determined Contributions (NDCs) under the Paris Agreement, providing financial incentives, technical assistance, and capacity-building support for farmers transitioning to regenerative farming systems.

Innovations in Agro-Technology

In the quest for sustainable agriculture, integrating innovations in agro-technology plays a crucial role. These innovations include tools for precision agriculture, soil health monitoring, and efficient water use. Here are some strategic recommendations for effectively integrating these technologies into farming practices.

Precision Agriculture

Cheering farmers to adopt precision farming techniques involves utilizing satellite imagery, drones, GPS technology, and sensor-based devices to optimize inputs such as fertilizers, pesticides, and water. Investment in data analytics platforms is promoted to process and analyze large datasets collected from various sources, including weather stations, soil sensors, and crop monitoring devices, to provide real-time insights and decision support to farmers. Training programs and resources are provided to farmers on the use of precision agriculture tools, as well as the interpretation of generated data, to maximize their effectiveness in optimizing farm management practices.

Soil Health Monitoring

To ensure the deployment of soil sensors and monitoring devices that can assess key soil parameters such as moisture levels, nutrient content, pH, and microbial activity in real time, integrate remote sensing technologies such as satellite imagery and aerial drones to monitor soil health indicators over large agricultural areas, enabling early detection of soil degradation and targeted intervention. Promote soil conservation practices such as cover cropping, minimum tillage, and organic amendments based on the insights provided by soil health monitoring technologies.

Efficient Water Use

Encouraging farmers to adopt advanced irrigation management systems, utilizing soil moisture sensors, weather data, and predictive modeling, is highly recommended to optimize water application and minimize wastage. These systems monitor soil moisture levels and weather conditions, allowing precise irrigation scheduling, thus reducing water usage while maintaining crop productivity. Promotion of water-saving techniques such as drip irrigation, mulching, and rainwater harvesting further contributes to water conservation. Drip irrigation delivers water directly to plant roots, minimizing evaporation and runoff, while mulching conserves soil moisture and suppresses weed growth. Rainwater harvesting enables farmers to capture and store rainwater for later use, reducing reliance on groundwater and surface water sources. Investment in research and development efforts aimed at developing drought-resistant crop varieties and innovative water-saving technologies tailored to specific agroclimatic conditions is crucial. Developing crops that require less water to thrive and implementing technologies like precision irrigation and soil moisture management systems can help farmers adapt to changing climate conditions and water availability challenges. In summary, encouraging the adoption of advanced irrigation management systems, promoting water-saving techniques, and investing in research and development efforts are essential strategies for sustainable water management in agriculture, enhancing productivity and resilience to water-related challenges.

Policy Support and Financial Incentives

To incentivize the adoption of innovative agro-technologies, policymakers should develop supportive policy frameworks. This can be achieved through subsidies, tax incentives, and regulatory measures that promote sustainable farming practices. Additionally, financial support and grants should be provided to farmers to make precision agriculture equipment, soil monitoring devices, and water-saving technologies more accessible and affordable. Fostering collaborations between government agencies, research institutions, technology providers, and private sector stakeholders is crucial to facilitate knowledge transfer, technology diffusion, and capacity building in agro-technology innovation.

Farmer Outreach and Extension Services

To strengthen extension services, provide farmers with information, training, and technical assistance on adopting and utilizing innovative agro-technologies. Establish demonstration farms or field trials to showcase the benefits and feasibility of precision agriculture, soil health monitoring, and efficient water-use practices, encouraging broader adoption among farmers. Create online platforms, workshops, and knowledge-sharing networks where farmers can exchange experiences, best practices, and success stories related to agro-technology innovation.

Community Engagement

Effective community engagement is vital for the success of any project or initiative, especially in the realm of sustainability and development. Involving local communities and stakeholders in planning and decision-making processes ensures that projects are more relevant, effective, and sustainable in the long term. Here, we outline strategic recommendations for effective community engagement, emphasizing its importance and providing actionable steps for implementation.

Importance of Community Engagement

Involving local communities in planning processes allows for the integration of their valuable knowledge about their environment, culture, and needs into decision-making, leading to the development of more informed and contextually appropriate solutions.

Ownership and Empowerment

Engaging communities empowers them to take ownership of projects and initiatives. When communities participate in decision-making, they are more likely to invest in the outcomes and commit to their success. Community engagement establishes trust between project implementers and local stakeholders. It fosters positive relationships based on mutual respect, transparency, and collaboration, which are essential for the success of any development endeavor. Projects that consider local values, traditions, and cultural practices are more likely to gain acceptance from communities. Engagement ensures that interventions are culturally sensitive and respectful of community norms and customs. Engaged communities are more likely to support and sustain projects beyond their initial implementation. By involving stakeholders in planning and decision-making, projects are better positioned to address long-term challenges and achieve lasting impact.

Strategic Recommendations

Engage with communities as early as possible in the project planning phase and maintain regular communication throughout the project life cycle. This allows for identifying community needs, concerns, and priorities from the outset. Recognize that different communities have unique characteristics, needs, and preferences.

Tailor engagement approaches to suit the specific context, culture, and language of the target community. This may involve utilizing a variety of communication channels, such as community meetings, workshops, or digital platforms. Facilitate meaningful participation of community members in decision-making processes. Provide opportunities for stakeholders to voice their opinions, contribute ideas, and actively shape project plans and activities. Invest in building the capacity of local communities to participate effectively in decision-making processes. Provide training, education, and resources to empower community members to engage meaningfully and contribute to project outcomes. Ensure transparency by providing clear and accurate information about project goals, processes, and outcomes. Communicate opcnly about project timelines, budgets, and potential risks, and address any concerns or questions raised by the community. Establish mechanisms for ongoing feedback and evaluation to assess the effectiveness of community engagement strategies and project implementation. Use feedback to adapt and refine approaches as needed to better meet the needs and expectations of the community. Forge partnerships with local organizations, community leaders, and other stakeholders to strengthen engagement efforts. Collaborate with existing community structures and networks to leverage local expertise and resources.

Future Research Directions

Call for Longitudinal Studies on the Impact of Regenerative Practices on One Health Outcomes

Future research should focus on longitudinal studies to investigate the impact of regenerative agricultural practices on One Health outcomes. While current research has provided valuable insights into the benefits of regenerative agriculture for soil health, biodiversity, and human nutrition, there is a need for a deeper understanding of its broader implications. Longitudinal studies tracking the effects of regenerative practices over time are crucial for assessing their sustainability and scalability. These studies should aim to quantify the impact of regenerative agriculture on soil health parameters such as soil organic matter content, microbial diversity, and nutrient cycling dynamics. Additionally, studies should evaluate the influence of regenerative practices on plant health, including crop yield, resilience to pests and diseases, and nutritional quality. Furthermore, longitudinal studies should investigate the effects of regenerative agriculture on animal health and welfare within farming systems. This includes assessing the impact of holistic grazing management and integration of livestock on pasture health, animal nutrition, disease prevalence, and overall well-being. Understanding how regenerative practices influence livestock health is crucial for optimizing the symbiotic relationships between plants and animals in agroecosystems. Moreover, these studies should examine the socio-economic aspects of regenerative agriculture, including its impact on farmer livelihoods, community resilience, and food security. By analyzing the socio-economic implications of adopting regenerative practices, researchers can provide valuable insights into the feasibility and adoption barriers of sustainable farming methods. In conclusion, longitudinal studies on the impact of regenerative practices on One Health outcomes are essential for advancing our understanding of sustainable agriculture and ensuring the well-being of ecosystems, communities, and future generations.

Exploration of New Technologies for Enhancing Soil Health and Monitoring Ecological Changes

In the domain of regenerative agriculture, ongoing research into innovative technologies plays a pivotal role in deepening our comprehension of soil health and ecological dynamics. For future endeavors, a prime focus should be on exploring and developing new technologies to bolster soil health and monitor ecological shifts. One avenue for forthcoming investigation lies in advanced soil management techniques, such as precision agriculture and digital farming technologies, which hold promise in optimizing soil health by precisely managing inputs like irrigation, fertilizers, and soil amendments. Integration of sensor technologies, remote sensing, and geographic information systems (GIS) offers the potential for real-time data on soil properties, moisture levels, and nutrient status, enabling informed decision-making by farmers to enhance soil health and productivity. Moreover, there is a pressing need to delve into emerging techniques for soil restoration and regeneration, including research into innovative soil amendment materials like biochar, compost, and microbial inoculants, which can bolster soil fertility, structure, and resilience. Exploring novel approaches for carbon sequestration in soils, such as agroforestry and regenerative grazing practices, should also be prioritized to mitigate climate change impacts and promote soil health. Additionally, future research should focus on developing advanced monitoring tools to assess ecological changes in agricultural landscapes, utilizing high-resolution satellite imagery, drones, and machine learning algorithms to monitor vegetation dynamics, land-use changes, and biodiversity patterns. Collaboration across disciplines, including agronomy, soil science, ecology, engineering, and data science, is imperative to advance the field of soil health and ecological monitoring comprehensively. Leveraging expertise from diverse disciplines can lead to holistic approaches in tackling complex challenges related to soil degradation, biodiversity loss, and climate change resilience in agricultural systems. Ultimately, investing in research and innovation holds the promise of developing practical solutions to promote soil health, biodiversity, and ecosystem resilience, thereby contributing to the long-term sustainability of agricultural production systems.

Conclusion

Regenerative agriculture represents a promising paradigm shift in agricultural practices. By embracing its principles and scaling up implementation, we can create a more sustainable and resilient food system that benefits both people and the planet. However, realizing the full potential of regenerative agriculture requires concerted efforts from policymakers, practitioners, and researchers to overcome barriers and facilitate widespread adoption.

Our critical findings highlight the potential of regenerative agriculture practices to enhance soil organic carbon (SOC) levels, particularly in restoring soil health. Despite variations in experimental designs, empirical evidence consistently demonstrates increases in SOC associated with these practices. However, certain practices, such as the use of organic amendments like compost, may result in increased greenhouse gas emissions, which need to be mitigated. This suggests that while regenerative agriculture holds promise for soil restoration and climate change mitigation, careful consideration of potential emissions is crucial.

These findings have significant implications for global agriculture. They underscore the importance of transitioning towards regenerative agriculture practices to address soil degradation, enhance ecosystem resilience, and mitigate climate change. By adopting these practices on a larger scale, agricultural systems can become more sustainable, productive, and resilient to environmental challenges. Moreover, addressing greenhouse gas emissions from agricultural activities is critical for meeting global climate goals and ensuring food security for future generations.

Regenerative agriculture has the potential to serve as a transformative model for sustainable development. By prioritizing soil health, biodiversity, and ecosystem functioning, regenerative agriculture offers a holistic approach to agricultural production that aligns with the principles of One Health. It not only promotes environmental sustainability but also supports social and economic well-being by fostering resilient food systems and empowering local communities.

Furthermore, regenerative agriculture offers a pathway towards achieving multiple Sustainable Development Goals (SDGs), including those related to poverty alleviation, climate action, and biodiversity conservation. Its emphasis on natural processes, reduced chemical inputs, and ecosystem-based approaches can help address pressing global challenges while promoting long-term resilience and sustainability.

References

Abbas F, Hammad HM, Fahad S, Cerdà A, Rizwan M, Farhad W, Ehsan S, Bakhat HF (2017) Agroforestry: A sustainable environmental practice for carbon sequestration under the climate change scenarios—a review. Environ Sci Pollut Res 24:11177–11191

Alam MJ, Humphreys E, Sarkar MAR, Sudhir Y (2018) Comparison of dry seeded and puddled transplanted rainy season rice on the high Ganges River floodplain of Bangladesh. Eur J Agron 96:120–130

Alexanderson MS, Luke H, Lloyd DJ (2023) Regenerative farming as climate action. J Environ Manag 347:119063

Al-Kaisi MM, Lal R (2020) Aligning science and policy of regenerative agriculture. Soil Sci Soc Am J 84(6):1808–1820

Arunrat N, Sereenonchai S, Chaowiwat W, Wang C, Hatano R (2022) Carbon, nitrogen and water footprints of organic rice and conventional rice production over 4 years of cultivation: a case study in the lower north of Thailand. Agronomy 12. https://doi.org/10.3390/agronomy12020380

Bass AM, Bird MI, Kay G, Muirhead B (2016) Soil properties, greenhouse gas emissions and crop yield under compost, biochar and co-composted biochar in two tropical agronomic systems. Sci Total Environ 550:459–470

Beacham JD, Jackson P, Jaworski CC, Krzywoszynska A, Dicks LV (2023) Contextualising farmer perspectives on regenerative agriculture: a post-productivist future? J Rural Stud 102:103100

Bera T, Sharma S, Thind HS, Sidhu HS, Jat ML (2017) Soil biochemical changes at different wheat growth stages in response to conservation agriculture practices in a rice-wheat system of North-Western India. Soil Res 56:91–104

Bhaduri D, Sihi D, Bhowmik A, Verma BC, Munda S, Dari B (2022) A review on effective soil health bio-indicators for ecosystem restoration and sustainability. Front Microbiol 13:938481

Brown K, Schirmer J, Upton P (2021) Regenerative farming and human wellbeing: are subjective wellbeing measures useful indicators for sustainable farming systems? Environ Sustain Indic 11:100132

Brown K, Schirmer J, Upton P (2022) Can regenerative agriculture support successful adaptation to climate change and improved landscape health through building farmer self-efficacy and wellbeing? Curr Res Environ Sustain 4:100170

Buckton SJ, Fazey I, Sharpe B, Om ES, Doherty B, Ball P, Denby K, Bryant M, Lait R, Bridle S, Cain M (2023) The Regenerative Lens: a conceptual framework for regenerative social-ecological systems. One Earth 6(7):824–842

Bullock DG (1992) Crop rotation. Crit Rev Plant Sci 11(4):309–326

Cardoso A, Laviola BG, Santos GS, de Sousa HU, de Oliveira HB, Veras LC, Ciannella R, Favaro SP (2017) Opportunities and challenges for sustainable production of A. aculeata through agroforestry systems. Ind Crop Prod 107:573–580

Chabbi A, Lehmann J, Ciais P, Loescher HW, Cotrufo MF, Don A, SanClements M, Schipper L, Six J, Smith P et al (2017) Aligning agriculture and climate policy. Nat Clim Chang 7:307–309

Dash PK, Bhattacharyya P, Shahid M, Roy KS, Swain CK, Tripathi R, Nayak AK (2017) Low carbon resource conservation techniques for energy savings, carbon gain and lowering GHGs emission in lowland transplanted rice. Soil Tillage Res 174:45–57

de Bruyn LL, Jenkins A, Samson-Liebig S (2017) Lessons learnt: Sharing soil knowledge to improve land management and sustainable soil use. Soil Sci Soc Am J 81(3):427–438

Dey A, Dwivedi BS, Bhattacharyya R, Datta SP, Meena MC, Das TK, Singh VK (2016) Conservation agriculture in a rice-wheat cropping system on an alluvial soil of north-western Indo-Gangetic plains: Effect on soil carbon and nitrogen pools. J Indian Soc Soil Sci 64:246–254

Dey A, Ghosh A, Das S, Bhattacharyya R, Tigga P (2021) Belowground carbon storage and dynamics. In: Das S, Ghosh B, Das TK (eds) Soil science: fundamentals to recent advances. Springer, Cham, pp 49–67

Dhaliwal JK, Singh MJ, Sharma S, Gupta N, Kukal SS (2020) Medium-term impact of tillage and residue retention on soil physical and biological properties in dry-seeded rice–wheat system in north-West India. Soil Res 58:468–477

Evans MC, Carwardine J, Fensham RJ, Butler DW, Wilson KA, Possingham HP, Martin TG (2015) Carbon farming via assisted natural regeneration as a cost-effective mechanism for restoring biodiversity in agricultural landscapes. Environ Sci Pol 50:114–129

Giller KE, Hijbeek R, Andersson JA, Sumberg J (2021) Regenerative agriculture: an agronomic perspective. Outlook Agric 50(1):13–25

Gordon E, Davila F, Riedy C (2022) Transforming landscapes and mindscapes through regenerative agriculture. Agric Hum Values 39(2):809–826

Gordon E, Davila F, Riedy C (2023) Regenerative agriculture: a potentially transformative storyline shared by nine discourses. Sustain Sci 18(4):1833–1849
Grelet G, Lang S, Merfield C, Calhoun N, Robson-Williams M, Horrocks A, Dewes A, Clifford A, Stevenson B, Saunders CM, Lister C (2021) Regenerative agriculture in Aotearoa New Zealand–research pathways to build science-based evidence and national narratives. Environ Sustain Indic 11:100132
Hok L, Sa J, Boulakia S, Reyes M, Ferreira A, Tivet F, Saab S, Auccaise R, Inagaki T, Schimiguel R, Ferreira L, Briedis C, Canalli L, Kong R, Leng V (2021) Dynamics of soil aggregate-associated organic carbon based on diversity and high biomass-C input under conservation agriculture in a savanna ecosystem in Cambodia. Catena:198. https://doi.org/10.1016/j.catena.2020.105065
Janz B, Weller S, Kraus D, Racela H, Wassmann R, Butterbach-Bahl K, Kiese R (2019) Greenhouse gas footprint of diversifying rice cropping systems: Impacts of water regime and organic amendments. Agric Ecosyst Environ 270:41–54. https://doi.org/10.1016/j.agee.2018.10.011
Jat HS, Datta A, Sharma PC, Kumar V, Yadav AK, Choudhary M, Choudhary V, Gathala MK, Sharma DK, Jat ML, Yaduvanshi NPS (2018) Assessing soil properties and nutrient availability under conservation agriculture practices in a reclaimed sodic soil in cereal-based systems of North-West India. Arch Agron Soil Sci 64:531–545
Jat HS, Datta A, Choudhary M, Sharma PC, Yadav AK, Choudhary V, Gathala MK, Jat ML, McDonald A (2019a) Climate smart agriculture practices improve soil organic carbon pools, biological properties and crop productivity in cereal-based systems of North-West India. Catena 181:104059
Jat HS, Datta A, Choudhary M, Yadav AK, Choudhary V, Sharma PC, Gathala MK, Jat ML, McDonald A (2019b) Effects of tillage, crop establishment and diversification on soil organic carbon, aggregation, aggregate associated carbon and productivity in cereal systems of semi-arid Northwest India. Soil Tillage Res 190:128–138
Jayasinghe SL, Thomas DT, Anderson JP, Chen C, Macdonald BC (2023) Global application of regenerative agriculture: a review of definitions and assessment approaches. Sustainability 15(22):15941
Jian J, Du X, Reiter MS, Stewart RD (2020) A meta-analysis of global cropland soil carbon changes due to cover cropping. Soil Biol Biochem 143:107735
Jordon MW, Smith P, Long PR, Bürkner PC, Petrokofsky G, Willis KJ (2022) Can Regenerative Agriculture increase national soil carbon stocks? Simulated country-scale adoption of reduced tillage, cover cropping, and ley-arable integration using RothC. Sci Total Environ 825:153955
Jose S (2009) Agroforestry for ecosystem services and environmental benefits: an overview. Agrofor Syst 76:1–10
Kaye JP, Quemada M (2017) Using cover crops to mitigate and adapt to climate change. A review. Agron Sustain Dev 37:1–17
Khangura R, Ferris D, Wagg C, Bowyer J (2023) Regenerative agriculture—A literature review on the practices and mechanisms used to improve soil health. Sustainability 15(3):2338
Kibblewhite MG, Ritz K, Swift MJ (2008) Soil health in agricultural systems. Philos Trans R Soc B Biol Sci 363(1492):685–701
King AE, Blesh J (2018) Crop rotations for increased soil carbon: perenniality as a guiding principle. Ecol Appl 28(1):249–261
Kremen C, Miles A (2012) Ecosystem services in biologically diversified versus conventional farming systems: benefits, externalities, and trade-offs. Ecol Soc 17(4):40
Kumar N, Nath CP (2019) Impact of zero-till residue management and crop diversification with legumes on soil aggregation and carbon sequestration. Soil Tillage Res 189:158–167
Lal R (2006) Enhancing crop yields in the developing countries through restoration of the soil organic carbon pool in agricultural lands. Land Degrad Dev 17(2):197–209
Lal R (2015a) Soil health and carbon management. Food Energy Secur 4(2):117–131
Lal R (2015b) Restoring soil quality to mitigate soil degradation. Sustainability 7(5):5875–5895
Lienhard P, Tivet F, Chabanne A, Dequiedt S, Lelievre M, Sayphoummie S, Leudphanane B, Prévost-Bouré NC, Séguy L, Maron P-A, Ranjard L (2013) No-till and cover crops shift soil

microbial abundance and diversity in Laos tropical grasslands. Agron Sustain Dev 33:375–384. https://doi.org/10.1007/s13593-012-0099-4
McCauley K, Barlow K (2023) Regenerative agriculture: increasing plant diversity and soil carbon sequestration on agricultural landscapes. Surg J 15(1). https://doi.org/10.21083/surg.v15i1.7196
McClelland SC, Paustian K, Schipanski ME (2021) Management of cover crops in temperate climates influences soil organic carbon stocks: a meta-analysis. Ecol Appl 31(3):e02278
Meena RS, Singh AK, Jatav SS, Rai S, Pradhan G, Kumar S, Mina KK, Jhariya MK (2024) Significance of soil organic carbon for regenerative agriculture and ecosystem services. In: Biodiversity and bioeconomy. Amsterdam, Elsevier, pp 217–240
Mishra AK, Sinha DD, Grover D, Roohi, Mishra S, Tyagi R, Sheoran HS, Sharma S (2022) Regenerative agriculture as climate smart solution to improve soil health and crop productivity thereby catalysing farmers' livelihood and sustainability. In: Towards sustainable natural resources: monitoring and managing ecosystem biodiversity. Springer, Cham, pp 295–309
Mishra AK, Bhowmick MK, Peramaiyan P, Sharma S, Singh S (2024) Regenerative agriculture practices for rice-based systems in South Asia. In: Regenerative agriculture. CRC Press, Boca Raton, FL, pp 210–226
Miura T, Niswati A, Swibawa IG, Haryani S, Gunito H, Kaneko N (2013) No tillage and bagasse mulching alter fungal biomass and community structure during decomposition of sugarcane leaf litter in Lampung Province, Sumatra, Indonesia. Soil Biol Biochem 58:27–35
Muchane MN, Sileshi GW, Gripenberg S, Jonsson M, Pumariño L, Barrios E (2020) Agroforestry boosts soil health in the humid and sub-humid tropics: a meta-analysis. Agric Ecosyst Environ 295:106899
Nath CP, Das TK, Rana KS, Pathak H, Bhattacharyya R, Paul S, Meena MC (2017) Greenhouse gases emission, soil organic carbon and wheat yield as affected by tillage systems and nitrogen management practices. Arch Agron Soil Sci 63:1644–1660
Newton P, Civita N, Frankel-Goldwater L, Bartel K, Johns C (2020) What is regenerative agriculture? A review of scholar and practitioner definitions based on processes and outcomes. Front Sustain Food Syst 4:577723
O'Donoghue T, Minasny B, McBratney A (2024) Digital regenerative agriculture. npj Sustain Agric 2(1):5
Pandey A, Mai V, Vu D, Bui T, Mai T, Jensen L, de Neergaard A (2014) Organic matter and water management strategies to reduce methane and nitrous oxide emissions from rice paddies in Vietnam. Agric Ecosyst Environ 196:137–146. https://doi.org/10.1016/j.agee.2014.06.010
Parihar CM, Yadav MR, Jat SL, Singh AK, Kumar B, Yadvinder-Singh, Pradhan S, Chakraborty D, Jat ML, Jat RK, Saharawat YS, Yadav OP (2016) Long term effect of conservation agriculture in maize rotations on total organic carbon, physical and biological properties of a sandy loam soil in north-western Indo-Gangetic Plains. Soil Tillage Res 161:116–128
Parihar CM, Parihar MD, Sapkota TB, Nanwal RK, Singh AK, Jat SL, Nayak HS, Mahala DM, Singh LK, Kakraliya SK, Stirling CM (2018) Long-term impact of conservation agriculture and diversified maize rotations on carbon pools and stocks, mineral nitrogen fractions and nitrous oxide fluxes in Inceptisol of India. Sci Total Environ 640:1382–1392
Parihar CM, Singh AK, Jat SL, Ghosh A, Dey A, Nayak HS, Parihar MD, Mahala DM, Yadav RK, Rai V, Satayanaryana T et al (2019) Dependence of temperature sensitivity of soil organic carbon decomposition on nutrient management options under conservation agriculture in a sub-tropical Inceptisol. Soil Tillage Res 190:50–60
Parihar CM, Singh AK, Jat SL, Dey A, Nayak HS, Mandal BN, Saharawat YS, Jat ML, Yadav OP (2020) Soil quality and carbon sequestration under conservation agriculture with balanced nutrition in intensive cereal-based system. Soil Tillage Res 202:104653
Park MS, Baral H, Shin S (2022) Systematic approach to agroforestry policies and practices in Asia. Forests 13:635. https://doi.org/10.3390/f13050635
Pauli N, Abbot LK, Negrete-Yankelevich S, Andres P (2016) Farmers' knowledge and use of soil fauna in agriculture: a worldwide review. Ecol Soc 21(3):19. https://doi.org/10.5751/ES-08561-210319

Pörtner HO, Roberts DC, Poloczanska ES, Mintenbeck K, Tignor M, Alegría A, Craig M, Langsdorf S, Löschke S, Möller V, Okem A (2022) IPCC, 2022: Summary for policymakers
Rehberger E, West PC, Spillane C, McKeown PC (2023) What climate and environmental benefits of regenerative agriculture practices? An evidence review. Environ Res Commun. 5:052001
Richelle L, Visser M, Bock L, Walpole P, Mialhe F, Colinet G (2018) Looking for a dialogue between farmers and scientific soil knowledge: learnings from an ethno-geomorphopedological study in a Philippine's upland village. Agroecol Sustain Food Syst 42(1):2–27. https://doi.org/10.1080/21683565.2017.1322661
Sahu G, Das S (2020) Regenerative agriculture: future of sustainable food production. Biotica Res Today 2(8):745–748
Sanchis E, Ferrer M, Torres AG, Cambra-López M, Calvet S (2012) Effect of water and straw management practices on methane emissions from rice fields: a review through a meta-analysis. Environ Eng Sci 29:1053–1062. https://doi.org/10.1089/ees.2012.0006
Schreefel L, Schulte RP, De Boer IJM, Schrijver AP, Van Zanten HHE (2020) Regenerative agriculture–the soil is the base. Glob Food Sec 26:100404
Shah KK, Modi B, Pandey HP, Subedi A, Aryal G, Pandey M, Shrestha J (2021) Diversified crop rotation: an approach for sustainable agriculture production. Adv Agric 2021:1–9
Sharma S, Saikia R, Thind HS, Singh Y, Jat ML (2021) Tillage, green manure and residue management accelerate soil carbon pools and hydrolytic enzymatic activities for conservation agriculture based rice-wheat systems. Commun Soil Sci Plant Anal 52(4):470–486
Shin S, Soe KT, Lee H, Kim TH, Lee S, Park MS (2020) A systematic map of agroforestry research focusing on ecosystem services in the Asia-Pacific Region. Forests 11:368. https://doi.org/10.3390/f11040368
Singh VK, Yadvinder-Singh, Dwivedi BS, Singh SK, Majumdar K, Jat ML, Mishra RP, Rani M (2016) Soil physical properties, yield trends and economics after five years of conservation agriculture based rice-maize system in North-Western India. Soil Tillage Res 155:133–148
Singh I, Hussain M, Manjunath G, Chandra N, Ravikanth G (2023) Regenerative agriculture augments bacterial community structure for healthier soil and agriculture. Front Agron 5:1134514
Somasundaram J, Salikram M, Sinha NK, Mohanty M, Chaudhary RS, Dalal RC, Mitra NG, Blaise D, Coumar MV, Hati KM, Thakur JK (2019) Conservation agriculture effects on soil properties and crop productivity in a semiarid region of India. Soil Res 57:187–199
Song K, Zheng X, Lv W, Qin Q, Sun L, Zhang H, Xue Y (2019) Effects of tillage and straw return on water-stable aggregates, carbon stabilization and crop yield in an estuarine alluvial soil. Sci Rep 9(1):4586
Tahat MM, Alananbeh KM, Othman YA, Leskovar DI (2020) Soil health and sustainable agriculture. Sustainability 12(12):4859
Vamshi M, Jagadeesan R, Lamani HD, Rout S, VijayKumar R, Jagadesh M, Sachan K (2024) The revolutionary impact of regenerative agriculture on ecosystem restoration and land vitality: a review. J Geogr Environ Earth Sci Int 28(4):1–14
Vinke K, Martin MA, Adams S, Baarsch F, Bondeau A, Coumou D, Donner RV, Menon A, Perrette M, Rehfeld K, Robinson A et al (2017) Climatic risks and impacts in South Asia: extremes of water scarcity and excess. Reg Environ Chang 17(6):1569–1583
White KE, Brennan EB, Cavigelli MA, Smith RF (2020) Winter cover crops increase readily decomposable soil carbon, but compost drives total soil carbon during eight years of intensive, organic vegetable production in California. PLoS One 15(2):e0228677
Wurz A, Tscharntke T, Martin DA, Osen K, Rakotomalala AANA, Raveloaritiana E, Andrianisaina F, Dröge S, Fulgence TR, Soazafy MR et al (2022) Win-win opportunities combining high yields with high multi-taxa biodiversity in tropical agroforestry. Nat Commun 13:4127. https://doi.org/10.1038/s41467-022-30866-8
Yadav GS, Babu S, Das A, Bhowmik SN, Datta M, Singh R (2019) Soil carbon dynamics and productivity of rice–rice system under conservation tillage in submerged and unsubmerged ecologies of eastern Indian Himalaya. Carbon Manag 10:51–62
Yu Z, Lu C, Hennessy DA, Feng H, Tian H (2020) Impacts of tillage practices on soil carbon stocks in the US corn-soybean cropping system during 1998 to 2016. Environ Res Lett 15(1):014008

Yu T, Mahe L, Li Y, Wei X, Deng X, Zhang D (2022) Benefits of crop rotation on climate resilience and its prospects in China. Agronomy 12(2):436

Zazo-Moratalla A, Troncoso-Gonźalez I, Moreira-Mũnoz A (2019) Regenerative food systems to restore urban-rural relationships: Insights from the concepci'on metropolitan area foodshed (Chile). Sustainability 11. https://doi.org/10.3390/su11102892

Zomer RJ, Neufeldt H, Xu J, Ahrends A, Bossio D, Trabucco A, Van Noordwijk M, Wang M (2016) Global Tree Cover and Biomass Carbon on Agricultural Land: the contribution of agroforestry to global and national carbon budgets. Sci Rep 6(1):29987

Optimizing Soils for Regenerative Agriculture: A Case Study in Odisha

2

Antaryami Mishra, Ajay Kumar Mishra,
Piyush Kumar Maurya, and Sheetal Sharma

Introduction

Odisha, located in the eastern part of India, is a region characterized by a diverse range of geographical features and climatic conditions that significantly influence its agricultural practices. Spanning from 17°41′ to 22°33'N latitude and 81°21′ to 87°30′E longitude, Odisha covers an area of approximately 15.57 million hectares, making up about 4.7% of India's total geographical area. The state is bordered by West Bengal, Jharkhand, Andhra Pradesh, Chhattisgarh, and the Bay of Bengal.

The state experiences a tropical monsoon climate, marked by high humidity, medium to high rainfall, and moderate winters. The average annual rainfall is about 1497 mm, with about 80% received during the monsoon months from mid-June to mid-September. This distribution, however, is erratic, which poses challenges for agricultural reliability and water resource management. The temperature varies widely across the state with mean summer temperatures at around 30.3 °C, which can rise up to 46 °C, and mean winter temperatures at around 21.3 °C.

Agriculture is a crucial component of Odisha's economy, contributing approximately 28% of the net state domestic product and engaging around 64% of the workforce. Over 70% of the state's population depends directly on agriculture for

A. Mishra
Odisha University of Agriculture and Technology, Bhubaneswar, Odisha, India

Siksha O Anusandhan University, Bhubaneswar, Odisha, India

A. K. Mishra (✉) · P. K. Maurya · S. Sharma
International Rice Research Institute South Asia Regional Centre,
Varanasi, Uttar Pradesh, India
e-mail: a.k.mishra@irri.org

A. K. Mishra et al. (eds.), *Key Drivers and Indicators of Soil Health Management*, https://doi.org/10.1007/978-981-97-7564-4_2

their livelihood, underscoring the importance of land and soil management in ensuring economic stability and food security.

Odisha's land is a mosaic of diverse soil types, shaped by its complex topography and varied climatic zones. The state is divided into several physiographic regions, each with distinct soil characteristics that influence the types of crops that can be cultivated, the methods of cultivation, and the potential agricultural output. From the fertile soils of the coastal plains to the more challenging terrains of the Eastern Ghats and plateaus, the diversity in soil types includes red, black, lateritic, and deltaic alluvial soils, each requiring specific management practices to optimize agricultural productivity.

Regenerative agriculture (RA) represents a holistic approach to farming that seeks to rehabilitate and enhance the entire ecosystem of the farm. By focusing on soil health, RA practices aim to improve water retention, enhance biodiversity, increase resilience to climate variability, and reduce the reliance on chemical inputs such as fertilizers and pesticides (Grover et al. 2024; Mishra et al. 2024; Mishra et al. 2022c). These practices include, but are not limited to, cover cropping, no-till or reduced tillage, organic farming, crop rotation, and permaculture.

In the context of Odisha, RA holds significant promise due to its potential to improve soil fertility, reduce erosion, and mitigate the effects of climate change. By integrating these practices, farmers can potentially transform their lands into more productive and sustainable systems, enhancing both their yields and the health of their landscapes (Mishra et al. 2022a). Furthermore, regenerative practices can help stabilize the local climate, sequester carbon in the soil, and create more resilient agricultural ecosystems (Sahu and Mishra 2005; Mishra et al. 2011).

In a nutshell, Odisha's diverse and variable climatic and geographical conditions present unique challenges and opportunities in agriculture. Adopting RA principles can play a crucial role in addressing these challenges, promoting sustainable land use, and enhancing the well-being of the vast majority of the population that depends on agriculture for their livelihood.

As agriculture serves as the backbone of Odisha's economy, supporting the majority of its population, the chapter delves into the critical role that soil health plays in enhancing agricultural productivity and sustainability. By detailing the characteristics of various soil types across the state and their specific management strategies, the chapter offers valuable insights into how RA practices can be tailored to local conditions to improve soil fertility, reduce erosion, and increase resilience to climatic variations.

Furthermore, the chapter emphasizes the potential of RA to not only improve yields and livelihoods but also address broader environmental challenges such as carbon sequestration and biodiversity loss. This aligns with global sustainability goals, making it a crucial reference for policymakers, agricultural professionals, and researchers aiming to foster sustainable development through agriculture. The knowledge disseminated through this chapter can catalyze the transition towards more sustainable farming systems in Odisha and similar agrarian regions worldwide (Mishra et al. 2022b).

Soil Types and Distribution

Based on the geographic and geological features along with erosional processes, the state has been divided into seven physiographic regions, namely the Eastern Ghats, Garhjat Hills, Dandakaranya Region, Mahanadi basin, Bengal basin, Utkal Plain, and Midnapore delta. Each region is further divided into ten agroclimatic zones.

1. **The Eastern Ghats Region:** It consists of hill ranges covering 36% of the total area of the state. The region comprises the districts of Malkangiri, Koraput, Rayagada, Kandhamal, Kalahandi, and part of Gajapati and Ganjam. These are composed of khondalites, charnockites, and granite gneisses.
2. **The Garhjat Hills:** It occupies 26.4% of the total geographical area of the state covering the districts of Mayurbhanj, Keonjhar, Dhenkanal, Angul, Deogarh, Sambalpur, and Sundargarh. The Brahmani and Baitarani rivers dissect it into three divisions. The eastern region consists of the densely forested and hilly areas of Mayurbhanj district. The middle region occupies the plateaus and valleys of Keonjhar district and parts of Dhenkanal and Sundargarh districts. The western region is flat topped and steep-edged with forest cover. The region predominantly consists of rocks of middle proterozoic age.
3. **The Mahanadi Basin:** This region covers areas of Sambalpur, Jharsuguda, Bargarh, Subarnapur, Boudh, and Angul and Dhenkanal districts. In this region are observed isolated hills and thick river valleys abruptly from the plains. It includes the extensive middle part of Mahanadi valley and the lower Tel river basin. It covers 20.16% of the total geographical area of the state.
4. **The Dandakaranya Region:** It is a highly dissected plateau which is mostly formed of gneisses, charnockites, and khondalites. It covers 16.25% of the total geographical area of the state. It covers parts of the districts of Nabarangpur and Nuapada. The major portion of this region is drained by the tributaries of river Mahanadi, namely Tel, Jonk, Udanti, Hathi, and Sonbhadra.
5. **The Bengal Basin:** It lies in the Ganga delta occupying parts of Balasore and Mayurbhanj district and occupies 1.33% of the total geographical area of the state. The region is of alluvial deposits.
6. **The Utkal Plain:** It consists of the alluvium along with the patches of Balasore, Bhadrak, Puri, Khurda, and Ganjam districts along the Bay of Bengal. It occupies 7.25% of the total geographical area of the state. Many mangrove swamps are seen in this region.
7. **The Mahanadi Delta:** It has been formed by the deltaic sediment of Mahanadi, Brahmani, and other rivers in recent past. It occupies the districts of Kendrapara, Jagatsinghpur, Jajpur, Puri, and parts of Bhadrak, Cuttack, and Khurda. It occupies around 7.6% of the total geographical area of the state. This region is subjected to heavy flooding.

The above physiographic regions were however grouped into four physiographic regions by the early workers (Mishra 1972) of soil science in Odisha. These regions are:

1. The Northern Plateau
2. Central Table Land
3. Eastern Ghats
4. Coastal Plain

With regard to the integration of the effect of land form, topography, climate, soil, and crop adaptability, the state has been divided into ten agroclimatic zones, the details of which are given in the table presented in the following page.

As per the earlier system of soil classification, which is in vogue even at the present in view of its simplicity, Odisha has been divided into eight broad soil groups such as:

1. Red soil
2. Mixed red and yellow soil
3. Black soil
4. Laterite soil
5. Deltaic alluvial soil
6. Coastal saline and alluvial soil
7. Brown forest soil and mixed red and black soil

Discussions will be made in detail about these soil groups in the following pages.

As per "Soil Taxonomy," these eight broad groups of soils come under four orders such as (Mishra 1972; Sahu and Mishra 2005; Mishra et al. 2011):

1. Inceptisols (49%)
2. Alfisols (25%)
3. Entisols (10%)
4. Vertisols (6%)

These four orders are divided into ten suborders. The Inceptisols have (a) Aquepts, (b) Ustepts, and (c) Ochrepts suborders; the Alfisols have (a) Ustalfs and (b) Aqualfs suborders; the Entisols have (a) Orthents, (b) Aquents, (c) Fluvents, and (d) Psamments suborders; and the Vertisols have (a) Usterts suborder.

The ten suborders are again divided into 22 great-groups, the detailed discussion of which is beyond the scope of this book (Table 2.1).

The above taxonomic classification of soils has great bearing on the fertility status of soils of different districts of Odisha. Even the undulating topography starting from ridge to valley, which is of common occurrence in the northern and southern parts of the state with variation in pedogenesis, contributes to the great variation in fertility status within districts, within blocks, and even within different villages of the same block.

Table 2.1 Description of agroclimatic zones of Odisha

Sl. No.	Agroclimatic zone	District	Area (m.ha)	Mean annual rainfall (mm)	Mean max. summer temp (°C)	Mean min. winter temp (°C)	Climate	Broad soil groups
1	North Western Plateau	Sundargarh	1.2	1648	38.8	15.0	Hot and moist subhumid	Mixed red and yellow (Haplustepts, Paleustalfs, Ustochrepts)
2	North Central Plateau	Mayurbhanj, Keonjhar	1.72	1535	36.6	11.1	Hot and moist subhumid	Red, mixed red, and black (Association of Alfisols and Vertisols)
3	North Eastern Coastal Plain	Balasore, Cuttack, Jajpur, Bhadrak	0.95	1568	36.0	14.8	Moist subhumid	Coastal alluvial and saline (Haplaquepts, Haluaquents)
4	East and South Eastern Coastal Plain	Cuttack, Jagatsinghpur, Kendrapara, Banki, Puri, Ganjam	2.04	1449	39.0	11.5	Hot and humid	Deltaic alluvial, saline, and laterite (Haplaquepts, Fluvaquents, Ustochrepts, Haplustepts)
5	North-Eastern Ghats	Khurda, Nayagarh, Kandhamal, Boudh	2.85	1597	37.0	10.4	Hot and moist subhumid	Red loam, brown forest soils (Haplustepts, Rhodustalfs, Ustochrepts, Ustorthents)

(continued)

Table 2.1 (continued)

Sl. No.	Agroclimatic zone	District	Area (m.ha)	Mean annual rainfall (mm)	Mean max. summer temp (°C)	Mean min. winter temp (°C)	Climate	Broad soil groups
6	Eastern Ghats High Land	Koraput, Nabarangpur	0.95	1522	34.1	7.5	Warm and humid	Red laterite (Haplustepts, Rhodustalfs, Plinthustalfs, Ustorthents)
7	South Eastern Ghat	Koraput, Malkangiri, Rayagada, Gajapati	0.97	1522	34.1	13.2	Warm and Humid	Red, Black (Haplustepts, Rhodustalfs, Chromusterts, Ustorthents)
8	Western Undulating	Kalahandi, Nuapada, Padampur	1.27	1527	37.8	11.9	Hot and moist subhumid	Red, black (Haplustepts, Rhodustalfs, Chromusterts, Ustorthents)
9	West Central Table Land	Sambalpur, Bargarh, Bolangir, Sonepur	2.41	1527	40.0	12.4	Hot and moist subhumid	Lateritic, mixed red and yellow (Haplustepts, Paleustalfs, Ustochrepts)
10	Mid-Central Table Land	Dhenkanal, Angul	1.23	1421	38.7	14.0	Hot and dry subhumid	Red, lateritic, alluvial (Haplustepts, Plinthustalfs, Ochraqualfs)

Soil Types of Odisha and Their Physiochemical Characteristics

Each soil group (also referred to as soil type) is associated with specific characters and problems posing constraints for higher agricultural production (Fig. 2.1). These characters have been identified, and special attentions are made to increase productivity.

The characters of each soil along with their management practices are discussed below.

1. **Red soil:** Red soil covers about 7.14 million ha of land and is of the highest coverage among all soil groups of Odisha. It is found in the districts of Koraput, Rayagada, Nabarangpur, Malkangiri, Keonjhar, Ganjam, Kalahandi, Nuapada, Bolangir, Dhenkanal, and Mayurbhanj. Presence of excess amounts of oxidized iron imparts red color to the soil. The soils of the former four districts are heavier in texture, and the rest of the districts have light-textured soil. The soils have angular or subangular blocky structure. It is dominated by kaolinite and illite clay minerals. These soils are strongly to moderately acidic with low to medium organic carbon matter status and poor water-holding capacity.

   These soils are deficient in nitrogen and phosphorus. Micronutrients like boron and molybdenum are highly deficient in these soils. They have low cation-exchange capacity (CEC) and high phosphate and sulfur adsorption property. Soils are deficient in calcium and magnesium. Water-soluble phosphates get fixed and become nonavailable to crop plants. Application of insoluble phosphates 2 weeks before sowing seeds or mixed application of insoluble rock

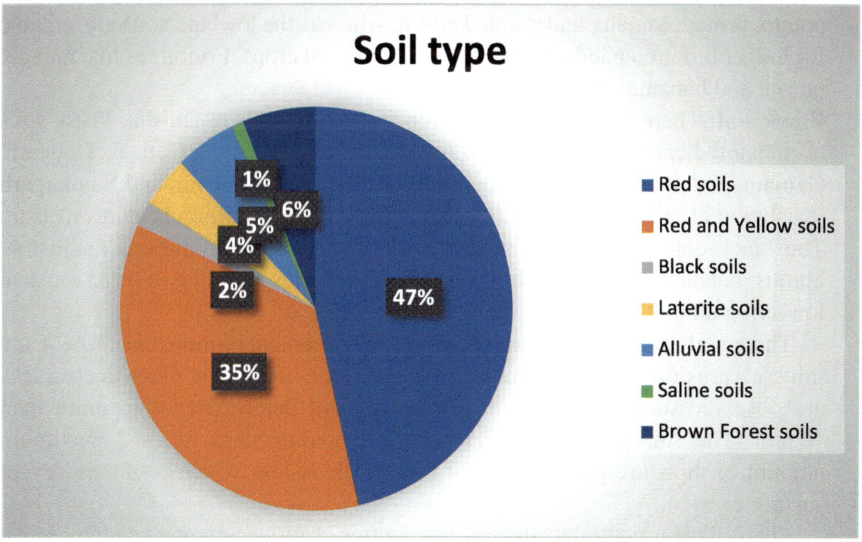

Fig. 2.1 Soil types of Odisha (classification as per earlier system)

phosphate, insoluble rock phosphate, and single superphosphate at equal proportion (1:1) makes the best utilization of phosphate.

Soil acidity is ameliorated by application of lime. Application of 5 quintals of papermill sludge (CPMS) every year has been advised for correction of soil acidity. Winter vegetables and groundnut crops among the application of 100 kg/ha crop treatment with sodium molybdate at a rate of 10 g/25 kg seeds correct molybdenum deficiency in pulses and groundnut crops. Application of 200 kg/ha of phosphogypsum meets the sulfur demand of oilseed crops. Crops like rice, finger millet, minor millets, ragi, potato, brinjal, and other vegetables, and fruit trees such as mango, jackfruit, guava, papaya, and sapota, are grown successfully in these soils.

2. **Mixed red and yellow soil:** These types of soils occupy the second highest area of the state spreading over 5.5 million ha of land. They are found in the districts of Sambalpur, Bargarh, Deogarh, and Sundargarh. They occur as catena associations in undulating and rolling topography which differs in depth, texture, and color. The soils are moderately shallow in depth and coarse textured. The upland soils are more shallow and lighter in texture than the lowland soils.

   Presence of ferruginous concretions and fluctuations in the water table imparts the mixed red and yellow colors to the soils. The upland soils are moderately acidic, whereas the lowland soils are slightly acidic. The lowland soils are formed mainly due to colluvial deposits.

   The upland soils are low in nitrogen and phosphate, whereas the lowland soils are medium in phosphate and high in potassium. Heavy light-textured soils are deficient in boron, and lowland soils with rice-rice cropping system under irrigated commands are deficient in zinc. Soil acidity can be corrected by the application of liming materials.

   The upland soils are suitable for small-duration rice, finger millet, sugarcane, potato, brinjal, tomato, and pointed gourd, whereas the lowland soils are suitable for longer duration paddy following pulses/oilseed crops. Fruit trees like mango, guava, and bananas grow well in these soils.

3. **Black soil:** There is no regular occurrence of black soils in Odisha. These soils occur sporadically in the districts of Angul, Bargarh, Bolangir, Cuttack, Deogarh, Ganjam, Kalahandi, Khurda, Malkangiri, Sambalpur, Sonepur, and Sundargarh (Mishra et al. 2011). These soils cover an area of 0.96 million hectares of land. The black color of the soil is due to the presence of titaniferous magnetite, humus, betaines, etc. These soils are formed due to weathering of basic rocks in low-lying areas.

   These soils are heavier in texture having clay content of more than 30%. Clay minerals are dominated with montmorillonite type of clay for which deep cracks are observed during summer. The effective soil depth extends to more than 90 cm. The soil provides very high water-holding capacity, and maximum amount of moisture permeability of these soils is low, which results in severe surface erosion.

   The soil pH is neutral to alkaline having free calcium carbonate nodules in the profile. The soils are rich in calcium but deficient in phosphorus, potassium, zinc, and boron. Upland rice suffers from iron deficiency in some places. Groundnut, mustard, and sunflower are found to respond to sulfur. Ammonia volatilization is high in paddy fields.

Soil moisture stress conditions are set early under drought conditions. Either in low- or high-moisture conditions, the soils cannot be ploughed. Management of these soils is difficult. Tillage operations should be completed at the right moisture consistency. Green manuring and application of bulky organic manures help in increasing water infiltration rates. Recycling of rice straw improves the aggregate formation of these soils. These soils are suitable for growing rice, jowar, bajra, maize, Bengal gram, safflower, mustard, and cotton.

4. **Laterite soils:** Lateritic soils are found in 0.70 million ha of land in the districts of Puri, Nayagarh, Cuttack, Dhenkanal, Keonjhar, Mayurbhanj, and Sambalpur. These soils are characterized by compact vesicular structure and are rich in hydrated oxides of iron and aluminum with small amounts of manganese, titanium, and quartz. Degraded laterites are loamy coarse-textured soils and are found in the districts of Khurda and Cuttack. These soils are loamy sand to sandy loam in the surface having clayey loam in the subsoil. Crusting is a problem in the upland laterite. Presence of a higher amount of exchangeable aluminum and manganese results in slightly acidic to strongly acidic soil with pH ranging between 4.5 and 5.8.

   Lateritic soils are poorly fertile with low organic matter. Available nitrogen and phosphorus are low, and potassium is medium. Nitrogen is lost due to leaching, and phosphates become unavailable due to fixation by Fe and Al oxides. Cation-exchange capacity (CEC) of the soil is low, and it is low in percentage base saturation (PBS). Sulfate is absorbed as pyrites or zinc sulfide.

   Medium- and lowland soils adjacent to uplands suffer from iron toxicity due to lateral movement of soluble iron from upper ridges. Crop productivity can be increased in these soils by proper soil management such as liming, application of organic manure, growing green manure crops, and applying balanced fertilizer with application of papermill sludge (CPMS) @ 5 quintals per hectare every year. Groundnut and pulses are grown successfully in these soils. Potato and other vegetables can also grow well. Application of a mixture of water-soluble and insoluble phosphatic fertilizers increases the crop yield. Soil application of boron and seed treatment with molybdenum improve legume yields. Rice, finger millets, minor millets, and sesame can grow well with proper manure and fertilizer application. Fruit trees like mango, jackfruit, banana, guava, and sapota grow well in lateritic soils.

5. **Deltaic alluvial soil:** These soils occupy 0.67 million ha of lands and occur in the deltaic regions of the rivers such as Mahanadi, Brahmani, Baitarani, Subarnarekha, and Rushikulya in the districts of Balasore, Bhadrak, Jajpur, Kendrapara, Jagatsinghpur, Cuttack, Puri, Ganjam, and Gajapati. Texture of these soils varies from coarse sand to clay, and it is mostly dependent on the geomorphology of the flood plains and the type of alluvial material transported by river water.

These soils are generally clay to clay loam in texture and columnar in structure. The pH of these soils varies between 6.6 and 8.0 with a conductivity of 10–45 dS/m in the summer. The exchangeable sodium percentage (ESP) varies between 18 and 27. The soils are rich in nitrogen and potassium and low to medium in phosphorus. The saline soils are sufficient in sulfate, boron, molybdenum, and chloride. Failure of crops normally occurs due to (i) plasmolysis of germinating seeds and roots; (ii) death of younger seedlings; (iii) reduced uptake of K, Ca, and Mg due to the presence of excess Na; (iv) toxicity due to B; and (v) hydrogen sulfide injury.

However, during the rainy season, the salinity hazards are low due to leaching and flushing of soluble salts by heavy rains. Rice is the main kharif crop with usual and well-distributed rainfall. Although kharif rice does not suffer very much from salinity, very often early drought and cyclonic seawater inundation cause hazards to rice. Salt-tolerant high-yielding rice varieties include Lunishree, SR-26B, and Maton, along with the local saline-resistant varieties like Sola, Pateni, and Cuttack Chandi. Salt-tolerant rabi crops such as safflower, mustard, barley, linseed, chili, sugar beet, tomato, spinach, and some cucumbers grow well in these soils. Cotton is a successful crop in saline soil if managed properly.

Coastal saline soils with high exchangeable sodium but that are acidic in reaction could be reclaimed by the application of lime, which replaces some of the exchangeable Na^+ or H^+ by Ca^{2+}. Use of organic manures, recycling of straw, and incorporation of green manured crops reduce the salinity and improve soil structure.

Cropping on alternate ridges of irrigated furrows with the intermediate furrow left as fallow and frequent light irrigation are some of the cultural practices that save crops from salinity. Preventive measures such as construction of salt embankments, provision of suitable drainage system, flush-out of soluble salts, construction of sluices across the creeks, raising shelter belts, and even any new saline belts check salinity hazards to a great extent.

6. **Brown forest soil:** Because of the association of this group of soil with forest areas of the state, this soil is called as brown forest soil. These are distributed in the districts of Kandhamal, Rayagada, and parts of Ganjam, Gajapati, and Nayagarh and cover about 0.17 million ha of land. These are brown to gray brown in color, light-colored texture, and acidic in reaction. Organic matter and nitrogen content of these soils are medium to high. Phosphorus and potash content are medium. The contents of most of the micronutrients in the soil are sufficient except molybdenum. In the slope terrains, soil erosion is heavy, making lands barren. In a few areas, shifting cultivation is a practice causing land degradation.

   Land shaping and, in some areas, podu cultivation can improve the soil and land management. With proper moisture conservation, these soils are most suitable for growing turmeric, ginger, and tropics. The organically produced turmeric of Kandhamal district is considered to have a reputation around the world. Maize, wheat, and mustard grow well in this soil. In the marginal land, niger is grown successfully with the use of nitrogen fertilizer. Horticultural crops like mango, jackfruit, guava, and citrus grow very well in these soils and have high commercial value. At present, social forestry plantations are being taken up successfully in the degraded soils.

7. **Mixed red and black soil:** These soils occur as associations of both red and black soil together in which black soil occurs in patches within the predominant red soil. The red and black soils are so intermixed that red soils are found in upper ridges whereas black soils occur in lower ridges. This soil group occupies about 0.16 million hectares of land in the western districts of Sambalpur, Bargarh, Sonepur, and Bolangir.

   Mixed red and black soils are light to medium textured having neutral pH. Black soils are rich in calcium, and red soils are dominant with iron, giving catena formation to the lands having these soils. The soils are generally deep with medium fertility status. The lowland soils growing rice are deficient in zinc. Rice, sugarcane, maize, ragi, groundnut, sesamum, and all types of vegetable crops are successfully grown in these soils with adequate fertilizer and organic manure applications.

   Furthermore, as per soil taxonomy classification, Inceptisols comprise the largest portion of Odisha soils, making up 49% of the total soil composition. Alfisols follow, accounting for 35%. Entisols represent 10%, while Vertisols are the least common, constituting 6% (Fig. 2.2). This distribution highlights the predominance of Inceptisols and Alfisols in the studied area. Such data is crucial for soil management and agricultural planning, as different soil orders have varying characteristics that influence crop suitability, water retention, and nutrient availability. Understanding soil composition aids in optimizing land use and enhancing agricultural productivity.

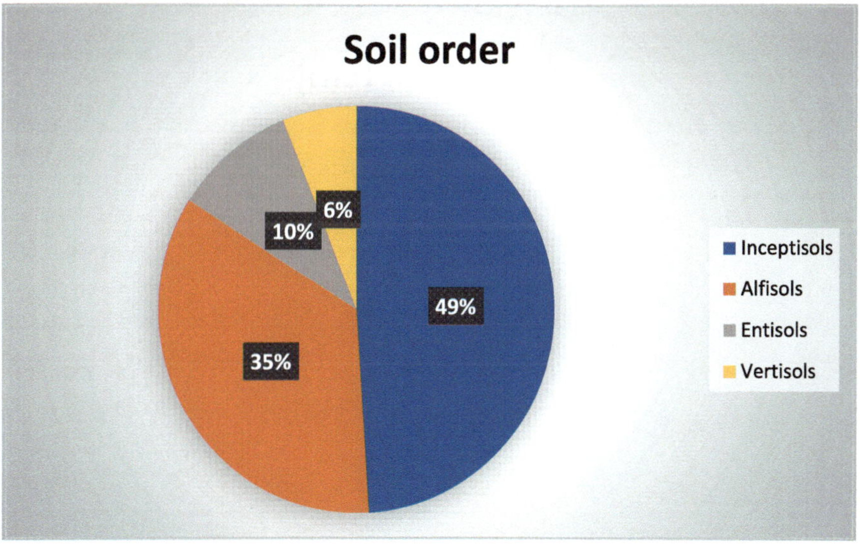

Fig. 2.2 Soil order of Odisha (classification based on soil taxonomy)

The pie chart in Fig. 2.3a represents the distribution of various soil textures. The largest segment is sandy loam, covering 48% of the total area. Clay loam follows with 28%, indicating significant clay content. Other soil textures include loamy sand (11%), sandy clay loam (6%), silty clay loam (4%), clay (2%), and loam (2%). This distribution highlights the predominance of sandy loam and clay loam in the surveyed area, suggesting that these soils are likely to influence agricultural practices and crop selection due to their specific physical properties and drainage characteristics.

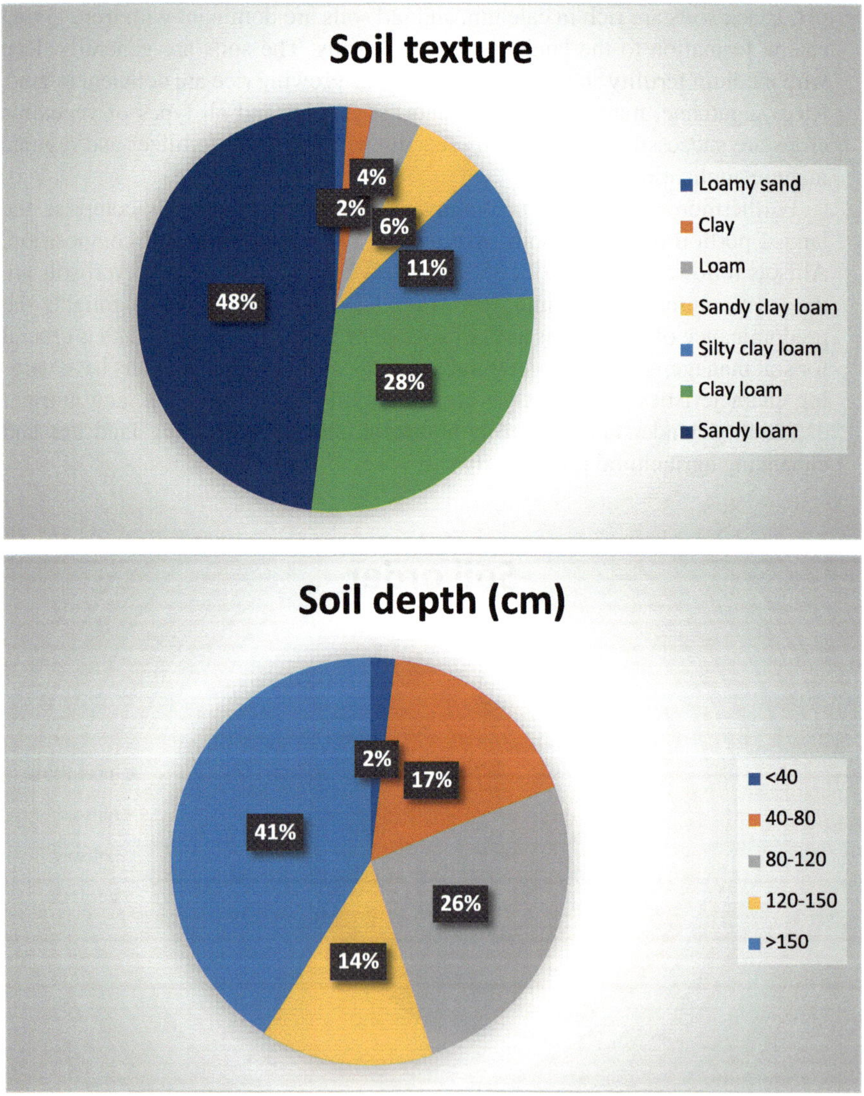

Fig. 2.3 (**a** and **b**) Soil texture distribution and soil depth variation

Figure 2.3b illustrates the variation in soil depth across different areas. The chart shows that 41% of the surveyed area has a soil depth of less than 40 cm, indicating shallow soil layers. A significant portion, 26%, has soil depths ranging from 80 to 120 cm, while 17% of the area has soil depths between 40 and 80 cm. Deeper soil layers, ranging from 120 to 150 cm, constitute 14% of the area, and only 2% have soil depths exceeding 150 cm. This variation in soil depth is crucial for determining root penetration, water retention capacity, and suitability of different crops for each soil depth category.

Soil Current Health Status of Odisha

Soil health is a combination of soil physical properties, soil chemical properties, and soil biological properties. Some mention has already been made on the physical-chemical properties of soils in the previous section while discussing various soil groups (soil types) of Odisha. A lot of work has not yet been done on the biological properties of Odisha soils covering all 30 districts except the organic carbon content of soils.

Previous research based on the analysis of 1,219,000 samples covering all the 30 districts of Odisha consisting of all the 314 blocks of the state reveals the following facts (Nanda et al. 2008):

As far as the soil reaction is concerned on a district-wise basis, 68% of the soils of Odisha are acidic, 6% are saline, and the rest 26% are neutral. Sixty percent of soils are medium, and 40% of soils are low in organic carbon. Seventy-three percent of soils are medium, and 27% of soils are low in available phosphorus. Concerning available potassium, 7% of soils are low, 68% are medium, and 7% are high. It is worth mentioning that as per this research of the year 2008, the districts of Boudh and Kalahandi are high, whereas the districts of Ganjam and Gajapati are low in available potassium content. Such variation in major available nutrient status among the various districts of the state and also among the various blocks within the districts has a direct bearing on the fertilizer recommendations for different crops in Odisha state.

A recent study (Dixit et al. 2020) was very extensive only on the same soil reaction, organic carbon, available phosphorus, potassium, sulfur, iron, manganese, copper, zinc, and boron. The findings of the study on concerned districts for all the 30 districts of the state are given in the following section.

Angul District

Sixty-six percent of soils of Angul district are acidic. Twenty-one percent of soils are neutral, and 13% of soils are saline. Thirty-four percent of soils of the district are deficient in organic carbon, 47% in available phosphorus, 11% in potassium, and 34% in sulfur. As far as micronutrients are concerned, 25% of soils are deficient in iron, 9% in manganese, 3% in copper, 61% in zinc, and 78% in boron.

Baleswar District

Eighty-six percent of soils of Baleswar district are acidic, 12% are neutral, and 2% are saline. Concerning major nutrients, 51% of soils are deficient in organic carbon, 47% in available phosphorus, 47% in potassium, and 38% in sulfur. Among the micronutrients, only 4% are deficient in iron, 6% in manganese, 3% in copper, 43% in zinc, and 70% in boron.

Bargarh District

The most agriculturally advanced district of the state is irrigated by one of the biggest irrigation commands of the country Hirakud Command Area; 79% of soils of the district are acidic, 15% are neutral, and 6% are saline. Thirty-seven percent of soils are deficient in organic carbon.

Known as the food bowl of the state, 36% of its soils are deficient in available phosphorus, 24% in potassium, and 28% in sulfur. As far as micronutrients are concerned, only 11% of soils are deficient in iron, 20% in manganese, 3% in copper, 54% in zinc, and 91% in boron.

Bhadrak District

Ninety-three percent of soils of this district are deficient in acidic reaction; 5% are neutral, and only 2% are saline. Forty-one percent of soils are deficient in organic carbon, 41% in available phosphorus, 25% in available potassium, and 50% in sulfur. Only 1% of soils are deficient in iron, 11% in manganese, 38% in zinc, and 72% in boron. All soils are sufficient in copper content.

Balangir District

Fifty-eight percent of the soils of Balangir district are acidic; 26% are neutral, and 16% are saline in reaction. Concerning major nutrients, 46% are deficient in organic carbon, 68% in available phosphorus, 16% in potassium, and 43% in sulfur. Among micronutrients, 6% of soils are deficient in iron, 7% in manganese, 5% in copper, 76% in zinc, and 92% in boron.

Boudh District

Fifty-eight percent of the soils of Boudh district are acidic in reaction; 29% are neutral, and 13% are saline. Fifty-one percent of soils are deficient in organic carbon, 50% in available phosphorus, 29% in potassium, and 52% in sulfur. Eight percent of soils are deficient in iron, 6% in manganese, only 2% in copper, 65% in zinc, and large amounts of soils to the extent of 93% in boron.

Cuttack District

Ninety percent of soils of Cuttack district are acidic, 7% are neutral, and 3% are saline. The soils of this alluvial bed of a bigger river of Mahanadi are agriculturally very important. Thirty eight percent of soils of this district are deficient in organic carbon, 37% in available phosphorus, 35% in potassium, and 45% in sulfur. Only 1% of soils are deficient in iron, 18% in zinc, and 85% in boron. All the soils are sufficient in manganese and copper.

Deogarh District

Deogarh district, literally meaning Abode of Divinity, which is also called an Ujjain of Odisha, is gifted by nature by mountains, springs, falls, forests, clear atmosphere, and rich collection of flora and fauna. The district's soils are acidic to the extent of 90%. Fourteen percent of soils are neutral, and 11% in saline. Twenty-nine percent of soils are deficient in organic carbon, 65% in available phosphorus, 74% in potassium, and 47% in sulfur. Only 2% of soils are deficient in iron, 1% in manganese, 11% in copper, 60% in zinc, and 80% in boron content.

Dhenkanal District

It is situated in the mid-central table land agroclimatic zone of the state. Eighty-three percent of soils of this district are acidic, 14% of soils are neutral, and 3% of soils are saline. Thirty-two percent of soils of the district are deficient in organic carbon, 59% in available phosphorus, 74% in potassium, and 42% in sulfur. Only 1% of soils are deficient in iron, 1% in manganese, 19% in zinc, and 76% in boron. All soils are sufficient in copper content.

Gajapati District

Ninety-three percent of soils of Gajapati district are acidic, 4% are neutral, and 3% are saline. Fifty-seven percent of soils are deficient in organic carbon, 50% in available phosphorus, 37% in potassium, and 71% in sulfur. Only 3% of soils are deficient in iron, 2% in manganese, 2% in copper, 47% in zinc, and 88% in boron.

Ganjam District

Seventy-one percent of soils of Ganjam district are acidic, 21% are neutral, and 8% are saline. Forty-seven percent of soils are deficient in organic carbon, 54% in available phosphorus, 17% in potassium, and 29% in sulfur. Only 1% of soils are deficient in iron, 3% in manganese, 31% in zinc, and 61% in boron. All soils of the district are sufficient in copper.

Jagatsinghpur District

Ninety-six percent of this coastal district soils are acidic, and 4% are neutral. Forty-five percent of soils are deficient in organic carbon, 31% in available phosphorus, 29% in potassium, and 59% in sulfur. Only 2% of soils are deficient in manganese, 56% in zinc, and 65% in boron. All soils of Jagatsinghpur district are sufficient in iron and copper content.

Jajpur District

Ninety-three percent of soils of this district are acidic, 5% are neutral, and 2% are saline. Forty-nine percent of soils are deficient in organic carbon, 33% in available phosphorus, 33% in available potassium, and 39% in sulfur. Thirty percent of soils are deficient in iron, 2% in manganese, 2% in copper, 26% in zinc, and 79% in boron content.

Jharsuguda District

Ninety-six percent of soils of this industrial dominated district are acidic, and 4% are neutral. Forty-four percent of soils are deficient in organic carbon, 55% in available phosphorus, 27% in available potassium, and 43% in sulfur. Only 2% of soils are deficient in manganese, 1% in copper, 32% in zinc, and 83% in boron. No soil is deficient in iron content.

Kalahandi District

Sixty-seven percent of soils of Kalahandi district are acidic, 27% neutral, and 12% saline. Forty-five percent of soils are deficient in organic carbon, 64% in available phosphorus, 10% in potassium, and 51% in sulfur. Only 3% of soils are deficient in iron, 4% in manganese, 11% in copper, 54% in zinc, and 80% in boron content.

Kendrapara District

Ninety-eight percent of soils of this district are acidic, and 2% are neutral. Thirty percent of soils are deficient in organic carbon, 42% in available phosphorus, 19% in available potassium, and 39% in sulfur. Twenty percent of soils of this coastal district are deficient in zinc and 42% in boron. All soils are sufficient in iron manganese and copper content.

Kandhamal District

Ninety-five percent of soils of this hilly forest-dominated district famous for organically grown turmeric cultivation are acidic, and 5% are neutral. Forty-one percent of soils are deficient in organic carbon, 50% in available phosphorus, 10% in available potassium, and 75% in sulfur.

Only 2% soils of Kandhamal district are deficient in iron, 6% in copper, 32% in zinc, and 86% in boron. No soil is deficient in manganese content.

Kendujhar District

Ninety percent of soils of Kendujhar district are acidic, 6% are neutral, and 2% are saline. Forty-six percent of soils are deficient in organic carbon, 37% in available phosphorus, 29% in available potassium, and 43% in sulfur. Two percent of soils are deficient in manganese, 20% in zinc, and 63% in boron. All soils are sufficient in iron content and copper content.

Khurda District

Ninety-two percent of soils of Khurda district are acidic, 7% are neutral, and 2% are saline. Sixty percent of soils are acidic and deficient in organic carbon, 48% in available phosphorus, 39% in available potassium, and 62% in sulfur. Only 2% of soils are deficient in manganese, 26% in zinc, and 81% in boron content. All soils are sufficient in iron and copper.

Koraput District

Ninety-five percent of soils of Koraput district are acidic, and 5% are neutral. Twenty-three percent of soils of this hilly terrain district of southern Odisha, which forms part of Eastern Ghats and Dandakaranya, are deficient in organic carbon, 42% are deficient in available phosphorus, 13% of soils are deficient in potassium, and 85% are deficient in sulfur. Forty-six percent of soils are deficient in zinc and 80% in boron content. All soils are sufficient in iron, manganese, and copper content.

Malkangiri District

Eighty-two percent of soils of Malkangiri district are acidic, 13% are neutral, and 5% are saline. Soils of this southern most district of Odisha, which also forms a part of Eastern Ghats mountain ranges and Dandakaranya, are deficient in organic carbon to the extent of 38% area. Sixty-eight percent of soils are deficient in available phosphorus, 29% in available potassium, and 51% in sulfur. Only 1% of soils are deficient in manganese, 2% in copper, 44% in zinc, and 85% in boron. All soils are sufficient in iron content.

Mayurbhanj District

Ninety-eight percent of soils of Mayurbhanj district are acidic, and 2% are neutral. Famous for Similipal Sanctuary and Similipal Tiger Project, 44% of soils of this hilly terrain district of north Odisha are deficient in organic carbon, 71% in available phosphorus, 46% in available potassium, and 67% in sulfur. Only 2% of soils are deficient in manganese, 11% in copper, 39% in zinc, and 95% in boron content. All soils are sufficient in iron.

Nabarangpur District

Coming under a part of Eastern Ghats mountain region and Dandakaranya region in southern Odisha, 91% of soils of this district are acidic, 8% are neutral, and 1% saline. This district is famous for maize cultivation in large area. Thirty-two percent of soils of Nabarangpur are deficient in organic carbon status, 61% in phosphorus content, 22% in potassium content, and 71% in sulfur content. Only 3% of soils are deficient in manganese, 1% in copper, 26% in zinc, and 89% in boron. All soils are sufficient in iron.

Nayagarh District

Seventy-six percent of soils of this district are acidic, 18% are neutral, and 6% are saline. Fifty-four percent of soils of this district are deficient in organic carbon, 39% in phosphorus, 29% in potassium, and 49% in sulfur. Only 1% of soils are deficient in manganese, 42% in zinc, and 81% in boron. All soils are sufficient in iron and copper.

Nuapada District

This district is situated in western Odisha and is adjacent to Chhattisgarh state. Forty-four percent of soils of Nuapada district are acidic, 33% are neutral, and 23% are saline. Twenty-two percent of soils are deficient in organic carbon, 65% in available phosphorus, only 1% in available potassium, and 26% in sulfur. Only 2% of soils are deficient in iron, 8% in manganese, only 1% in copper, 71% in zinc, and 80% in boron.

Puri District

Situated on the panoramic sea coast of the Bay of Bengal and being world famous for Shree Jagannath Temple, 95% of soils of this district are acidic, 3% are neutral, and 2% are saline. Forty-six percent of soils are deficient in organic carbon, 29% in available phosphorus, 37% in potassium, and 51% in sulfur. Eight percent of soils are deficient in manganese, 4% in copper, 30% in zinc, and 67% in boron. All soils are sufficient in iron content.

Rayagada District

Forming a part of Eastern Ghats mountain region, in south Odisha, 83% of soils of this district are acidic, 13% are neutral, and 4% are saline. Fifty-two percent of soils are deficient in organic carbon, 32% in available phosphorus, 61% in potassium, and 52% in sulfur. Only 4% of soils are deficient in iron, 11% in manganese, 2% in copper, 36% in zinc, and 76% in boron content.

Sambalpur District

Famous for the biggest dam of India—Hirakud Dam (25 km long)—over river Mahanadi and Maa Samaleswari Temple, 89% of soils of this district are acidic, 6% are neutral, and 5% are saline. Being irrigated by Hirakud Command Area, under the Intensive Agriculture District Programme (IADP), 41% of soils of this district are deficient in organic carbon, 52% in available phosphorus, 19% in available potassium, and 41% in sulfur. Only 2% of soils are deficient in iron, 3% in manganese, 2% in copper, 41% in zinc, and 96% in boron.

Subarnapur District

This district is known as the "Second Varanasi" of India for its cluster of temples (nearly 108) and also known as "Second Prayagraj" for the meeting point of three rivers, namely Mahanadi, Tel, and Suketi, in one place just like in Allahabad. Forty percent of the net sown area of this district is irrigated from all the sources as against 21.3% of the state average. This district stands second in average rice yield in the state after Jagatsinghpur district. Seventy percent of soils of this district are acidic, 21% are neutral, and 9% are saline. Thirty-seven percent of soils are deficient in organic carbon, 56% in available phosphorus, 25% in available potassium, and 23% in sulfur. Only 2% of soils are deficient in iron, 3% in manganese, 2% in copper, 41% in zinc, and 91% in boron content.

Sundargarh District

Situated in the northern part of Odisha adjacent to both Jharkhand and Chhattisgarh states, 94% of soils of this district are acidic, 5% are neutral, and only 1% are saline. Forty-nine percent of soils are deficient in organic carbon, 71% in available phosphorus, 17% in available potassium, and 47% in available sulfur. Two percent of soils are deficient in iron, 1% in manganese, 4% in copper, 72% in zinc, and 94% in boron.

Conclusion and the Way Forward

The study of Odisha's soils highlights the critical role that soil health plays in enhancing agricultural productivity and sustainability. The diverse range of soil types and their distinct characteristics necessitate tailored management practices to optimize their use. The adoption of regenerative agriculture (RA) practices, which focus on soil health, holds significant promise for improving soil fertility, reducing erosion, and mitigating climate change effects. These practices include cover cropping, no-till or reduced tillage, organic farming, crop rotation, and permaculture, which can transform lands into more productive and sustainable systems.

The findings from various districts underscore the widespread prevalence of acidic soils and deficiencies in essential nutrients like organic carbon, phosphorus, potassium, and sulfur, as well as micronutrients such as zinc and boron. Addressing these deficiencies through proper soil management practices, including the application of lime, organic manures, and balanced fertilizers, is crucial for enhancing crop productivity.

Way Forward

1. Adoption of regenerative agriculture practices: Encourage the widespread adoption of RA practices to improve soil health and sustainability. This includes promoting cover cropping, reduced tillage, and organic farming among farmers.
2. Soil health monitoring and management: Establish a robust soil health monitoring system to regularly assess soil conditions and nutrient levels across different regions. Provide farmers with customized soil management plans based on these assessments (Mishra et al. 2022a).
3. Nutrient management: Implement integrated nutrient management practices to address the deficiencies identified in the study. This includes the use of biofortified biochar, organic manures, and appropriate fertilizers to replenish nutrient levels.

4. Training and capacity building: Conduct training programs for farmers, extension workers, and agricultural professionals on the benefits and implementation of RA practices and effective soil management techniques.
5. Policy support and incentives: Advocate for policy support and financial incentives from the government to encourage the adoption of sustainable farming practices. This includes subsidies for organic fertilizers, biochar, and other soil amendments.
6. Research and development: Promote research on soil health and sustainable agriculture practices. Collaborate with academic institutions and research organizations to develop innovative solutions for soil management and crop productivity.
7. Climate resilience: Develop and promote climate-resilient farming practices that can withstand the variable climatic conditions of Odisha (Yadav et al. 2021). This includes selecting drought-resistant crop varieties and implementing water conservation techniques.

By integrating these strategies, Odisha can enhance its agricultural productivity, ensure food security, and promote sustainable development. The transition towards RA will not only benefit the local farming community but also contribute to broader environmental goals, such as carbon sequestration and biodiversity conservation.

References

Dixit S, Mishra PK, Muthukumaran M, Maheswar Reddy K, Pal PK, Poddar R, Padhee AK, Mishra A (2020) Mapping the nutrient status of Odisha's soils. ICRISAT, Hyderabad

Grover D, Mishra AK, Rani P, Kalonia N, Chaudhary A, Sharma S (2024) Soil management in sustainable agriculture: principles and techniques. In: Technological approaches for climate smart agriculture. Springer, Cham, pp 41–77

Mishra DP (1972) Agroclimatic Zones and Soils of Orissa. Directorate of Soil Conservation Govt. of Orissa

Mishra A, Dash BB, Das D (2011) Black soils of Orissa and their management. Orissa Rev. LXVII(1):24–26

Mishra AK, Yeggina PK, Sinha DD, Srivastava AK, Sharma S (2022a) Scope for production of pulses in rice fallow lands in South Asia. In: Advances in legumes for sustainable intensification. Academic Press, Cambridge, MA, pp 9–26

Mishra AK, Shinjo H, Jat HS, Jat ML, Jat RK, Funakawa S, Sutaliya JM (2022b) Farmers' perspectives as determinants for adoption of conservation agriculture practices in Indo-Gangetic Plains of India. Resour Conserv Recycling Adv 15:200105

Mishra AK, Sinha DD, Grover D, Roohi, Mishra S, Tyagi R, Sheoran HS, Sharma S (2022c) Regenerative agriculture as climate smart solution to improve soil health and crop productivity thereby catalysing farmers' livelihood and sustainability. In: Rani M, Chaudhary BS, Jamal S, Kumar P (eds) Towards sustainable natural resources. Springer, Cham. https://doi.org/10.1007/978-3-031-06443-2_16

Mishra AK, Bhowmick MK, Peramaiyan P, Sharma S, Singh S (2024) Regenerative agriculture practices for rice-based systems in South Asia. In: Rakshit A, Parihar M, Meena VS, Abhilash P, Jha PK, Sarkar D (eds) Regenerative agriculture: translating science to action, 1st edn. CRC Press, Boca Raton, FL, p 17. https://doi.org/10.1201/9781003309581

Nanda SK, Mishra A, Pradhan NK, Muralidharudu Y (2008) Soil Testing and Fertilizer Recommendation in Orissa. Technical Bulletin, AICRP on STCR, OUAT, Bhubaneswar

Sahu GC, Mishra A (2005) Soil of Orissa and its management. Orissa Rev LXI(4):56–60

Yadav P, Jaiswal DK, Sinha RK (2021) Climate change: impact on agricultural production and sustainable mitigation. In: Singh S, Singh P, Rangabhashiyam S, Srivastava KK (eds) Global climate change. Elsevier, Amsterdam, pp 151–174. https://doi.org/10.1016/B978-0-12-822928-6.00010-1

3 Assessment of Soil Microbial Communities: Methods and Their Impact on Soil-Plant Interactions

Laliteshwari Bhardwaj, Rajman Gupta,
and Prince Kumar Tiwari

Introduction

The soil microbiome is highly diverse, accounting for up to one-quarter of Earth's biodiversity (Wagg et al. 2019). Biological diversity is fundamental to ecosystem maintenance (Wagg et al. 2014). Soil biodiversity is crucial for sustaining multiple ecosystem functions, including nutrient cycling, organic matter decomposition, and plant productivity (Mishra et al. 2022; Feng et al. 2024). It enhances soil structure, thereby improving water infiltration and retention, and plays a vital role in disease suppression, carbon sequestration, and supporting plant growth. However, both anthropogenic factors such as intensive agriculture, deforestation, pollution, and urbanization and natural factors like climate change and extreme weather events pose significant risks by altering soil conditions and disrupting the balance of soil ecosystems (Tibbett et al. 2020).

Soil biodiversity is crucial for ecosystem functioning and provision of ecosystem services, encompassing a vast array of organisms such as bacteria, fungi, protozoa, nematodes, arthropods, and plant roots (Briones 2018). These diverse biological components drive essential ecological processes including nutrient cycling, organic matter decomposition, and soil formation (Kumar et al. 2022). Microorganisms are particularly vital in the mineralization of organic matter, converting it into plant-accessible

L. Bhardwaj (✉)
Department of Botany, Institute of Science, Banaras Hindu University,
Varanasi, Uttar Pradesh, India
e-mail: bharadwaj.93@bhu.ac.in

R. Gupta
School of Environmental Sciences, Jawaharlal Nehru University, New Delhi, India

P. K. Tiwari
Department of Mechanical Engineering, Indian Institute of Technology, BHU,
Varanasi, Uttar Pradesh, India

A. K. Mishra et al. (eds.), *Key Drivers and Indicators of Soil Health Management*, https://doi.org/10.1007/978-981-97-7564-4_3

forms, thereby enhancing soil fertility and plant productivity (Bhardwaj et al. 2024). Additionally, soil biota contributes to the stabilization of soil structure, improving its porosity and water retention capabilities, which in turn support plant growth and increase drought resilience (Coleman 2015; Mishra et al. 2022; Gupta et al. 2024). Moreover, a rich soil biota promotes disease suppression through competitive exclusion of pathogens and production of antimicrobial compounds (Mullis et al. 2019). The complex interactions between soil organisms and plants enhance plant health and productivity, forming the foundation of agricultural sustainability. Consequently, soil biodiversity is integral to maintaining ecosystem functions that ensure food security, climate regulation, water purification, and overall environmental health. Preserving and enhancing soil biodiversity are therefore essential for sustaining ecosystem services that are crucial for human well-being and environmental sustainability (Joshi and Garkoti, 2023; Reyes-Sánchez et al. 2022).

Soil biodiversity, which includes a diverse array of microorganisms, invertebrates, and plant roots, is integral to agricultural productivity, climate regulation, and environmental sustainability (Aytenew 2021). The diverse biotic components of soil ecosystems are pivotal in nutrient cycling processes, decomposing organic matter and making essential nutrients like nitrogen and phosphorus available, thereby enhancing crop yields and soil fertility (Bhardwaj et al. 2024). Microorganisms, which constitute the major biota in soil, are particularly crucial in land ecosystem restoration, and the ecological and economic benefits of the soil microbiome in agricultural practices are increasingly acknowledged (Kumar and Garkoti, 2022; Wang et al. 2022). Regarding climate regulation, soil biodiversity significantly contributes to carbon sequestration; soil organisms stabilize organic carbon within soil aggregates, thus reducing greenhouse gas emissions (Rattan et al. 2021). Furthermore, a rich soil biota improves soil structure and water retention, mitigating erosion and enhancing resilience to extreme weather events, which is vital for sustainable land management (Zhang et al. 2024). Therefore, preserving soil biodiversity is essential for maintaining ecosystem functions that support agricultural productivity, mitigate climate change effects, and ensure long-term environmental sustainability.

The role of soil biodiversity in regulating multiple ecosystem functions is not fully understood, which limits our ability to predict the consequences of soil biodiversity loss on ecosystem sustainability. This chapter aims to present a conceptual framework for assessing soil biodiversity and to explore the linkages between soil biodiversity and ecosystem functioning.

Methodological Approaches for the Assessment of Soil Microbial Diversity

Field Sampling Techniques

Various soil sampling techniques are utilized based on specific objectives: Core sampling involves extracting cylindrical soil sections for vertical analysis; grab sampling collects surface soil for contamination studies; composite sampling combines samples for averaged analysis over an area; profile sampling studies soil properties at different depths; auger sampling collects samples using soil augers for texture and chemical analysis; pit sampling excavates pits for comprehensive soil profile examination; and bulk density sampling measures soil mass per unit volume using core samplers for compaction and porosity assessment (Carter and Gregorich 2007; Rowell 2014; Weil and Brady 2016; Schoeneberger et al. 2012).

Sampling Design and Replication

Sampling design includes random sampling, systematic sampling, stratified sampling, and grid sampling. In random sampling, samples are collected across the study site to reduce biasness (Cochran 1977), while in systematic sampling, the samples are collected at regular intervals to ensure the even coverage of the study site (Thompson 2012). Stratified sampling divides the area into different strata, and sampling is done in each stratum. This method makes sure that all variations within the area are represented (Lohr 2021). The grid sampling method uses a grid layout over the study area and collects samples at intersections of the grid particularly useful for detailed spatial analysis (Webster and Oliver 2007). Replication is important for ensuring statistical validity and reliability of results. Generally, 3–5 replicates per sampling point are recommended to account for natural variability. Randomization of the sampling points and consistency in the sampling method should be maintained for effective results.

Diversity Assessment Through Dilutions and Culture-Dependent Techniques

Dilution Plating and Culturing Methods

Bacterial culture through selective and nonselective conditions allows researchers to isolate and study microorganisms in various contexts: selective conditions utilize specific inhibitors or nutrients to favor the growth of targeted bacteria, such as using MacConkey agar for Gram-negative bacteria, while nonselective conditions employ general-purpose media like nutrient agar to support a wide range of bacterial species, providing a comprehensive view of the microbial diversity present in a sample (Bonnet et al. 2019). Agar plate culturing is widely used for isolating soil microorganisms by spreading diluted soil samples onto nutrient, selective, or differential agar media to target diverse microbial groups (Eyler 2013). Enrichment culture is a technique used to increase the population density of a specific group of microorganisms within a sample's overall microbial population. This is accomplished by selectively promoting the growth of the target microorganisms through careful adjustment of physiological conditions during the enrichment phase (Kumar 2012; Power and Johnson 2009; Atlas 2010). The most probable number (MPN) method estimates the concentration of viable microorganisms by comparing growth patterns across

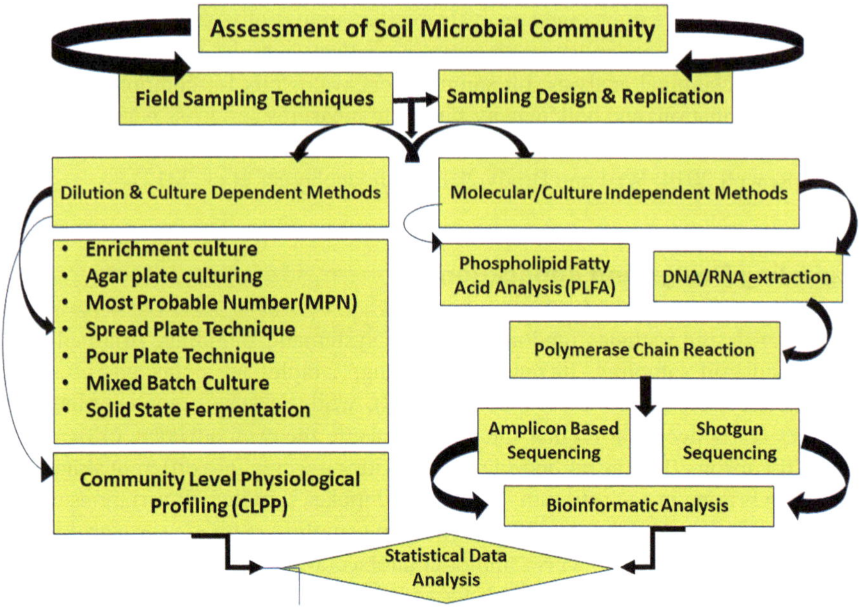

Fig. 3.1 The flowchart of soil microbial assessment

dilutions (Lares-Villa and Hernández-Peña 2010; Beattie et al. 2003). Spread plate and pour plate techniques enable counting and isolating aerobic and microaerophilic microbes (Terrones-Fernandez et al. 2023). Culturing microbial consortia in mixed-batch cultures or bioreactors reveals microbial interactions and metabolic capabilities. Solid-state fermentation studies fungi and actinomycetes on moist substrates (Yafetto 2022). The entire methods of study assessment have been summarized in Fig. 3.1.

Community-Level Physiological Profiles

Community-level physiological profiles (CLPPs) offer a reliable method to evaluate the functional diversity and metabolic potential of microbial communities by assessing their use of various carbon sources. This method typically involves the use of Biolog microplates, which are equipped with an array of carbon substrates that the microbial community from an environmental sample can metabolize (Perujo et al. 2020). When these samples are inoculated onto the microplates, microbial metabolic activity is detected through a color change due to a redox indicator dye (Garland, 1997). The resulting substrate utilization patterns, seen in these color changes, provide a detailed profile of the community's metabolic functions (Button et al. 2016). CLPPs are especially useful in fields like environmental monitoring, agriculture, waste management, and ecological studies, as they provide comprehensive insights into the functional potential of microbial communities and their responses to different environmental conditions or disturbances. Despite their benefits, such as simplicity and rapid execution, CLPPs primarily assess carbon utilization and may not fully represent the diversity of microbial communities, particularly non-culturable microorganisms (Jezierska-Tys et al. 2021). Furthermore, initial microbial biomass and activity levels can influence the results, necessitating careful data interpretation.

Culture-Independent Methods of Community Analysis

Phospholipid Fatty Acid Analysis

Phospholipid fatty acid (PLFA) analysis is a precise method for evaluating soil microbial communities by characterizing the specific fatty acids present in microbial cell membranes (Lewe et al. 2021). This technique involves extracting phospholipids from soil samples and subsequently identifying and quantifying the signature PLFAs associated with various microbial groups, including bacteria, fungi, and protozoa (Quideau et al. 2016). PLFA analysis is particularly advantageous for soil health assessments, as it is sensitive and capable of profiling both cultivable and non-culturable microorganisms, offering a comprehensive overview of microbial diversity and ecosystem functionality (Roberts et al. 2024).

DNA/RNA Extraction and Polymerase Chain Reactions

Polymerase chain reaction (PCR) amplifies specific DNA sequences from soil samples, allowing for the detection and quantification of targeted microbial groups. This technique is essential for identifying specific microorganisms, such as bacteria, archaea, and fungi, by targeting their unique genetic markers. Various types of PCR include conventional PCR (the basic form used to amplify DNA without providing quantitative data, involving repeated cycles of denaturation, annealing, and extension to generate multiple copies of a target DNA sequence); multiplex PCR which amplifies multiple targets simultaneously; nested PCR, utilizing two successive PCR reactions for higher specificity (Dubey et al. 2013); reverse transcriptase PCR (RT-PCR), converting mRNA to cDNA before amplification (Artika et al. 2022); quantitative PCR (qPCR), which quantifies DNA or RNA in a sample; hot-start PCR, reducing nonspecific amplification by heating reaction components before adding Taq polymerase (Luton et al. 2002; Fierer et al. 2005); touchdown PCR, involving a gradual decrease in annealing temperature to enhance specificity and efficiency (Green and Sambrook 2018); methylation-specific PCR (MSP), detecting DNA methylation patterns post-sodium bisulfite treatment (Li et al. 2011); and LAMP assay (loop-mediated isothermal amplification), amplifying DNA at a constant temperature using multiple primers and Bst polymerase (Wong et al. 2018).

High-Throughput Sequencing Techniques

Advances in metagenomic analysis, using next-generation sequencing, have significantly expanded our understanding of microbial communities compared to conventional techniques. These advances offer a modern approach to identifying novel and uncultivable microorganisms based on their genetic information from specific environments (Bhardwaj et al. 2023). Shotgun metagenomics involves analyzing the DNA of an entire community without the need for PCR amplification, allowing for the study of all genes present in a sample. Conversely, amplicon sequencing targets taxonomically important marker genes, limiting analysis to previously known DNA sequences. Sequence-based metagenomics involves directly analyzing DNA sequences from the environment without constructing a library, though it offers limited identification of novel genes and products. This can be supplemented by functional genomics. In contrast, function-based metagenomics requires the fragmentation and cloning of extracted metagenomic DNA into a suitable host, followed by functional screening and sequencing of clones to detect novel genes (Sharpton 2014; Quince et al. 2017; Lema et al. 2023). Literature survey on soil biodiversity studies across different ecosystems, highlighting the assessment methods and key findings, are listed in Tables 3.1 and 3.2.

Bioinformatics and Statistical Data Analysis

Bioinformatics is essential to life science research, playing a key role in microbial assessment by analyzing genomic, proteomic, and metagenomic data (Attwood et al. 2019). It assists in the identification and classification of microorganisms, gene annotation, and exploration of microbial diversity and community structure through methods such as sequence alignment, phylogenetic analysis, and metagenomic sequencing (Pérez-Cobas et al. 2020). Tools like BLAST, QIIME, and MEGA are used to process and interpret microbial data, providing insights into microbial ecology and evolution (Navas-Molina et al. 2013; Zamkovaya et al. 2021). When combined with statistical data analysis techniques such as regression analysis, cluster analysis, and machine learning, bioinformatics facilitates the extraction of meaningful insights from complex biological data (Morris and Baladandayuthapani 2017). This integration propels advancements in genomics, proteomics, systems biology, personalized medicine, and biotechnology, ultimately enhancing human health outcomes (Pinu et al. 2019).

Soil-Microbe-Plant Interactions

Soil-microbe-plant interactions constitute a fundamental component of terrestrial ecosystems, coordinating nutrient dynamics, plant physiology, and ecosystem stability (Chauhan et al. 2023). Microbial communities, spanning bacteria, fungi, and archaea, inhabit soil matrices in immense diversity and abundance, exerting

Table 3.1 National level literature survey on soil biodiversity studies across different ecosystems

S. No.	Soil biodiversity studied	Ecosystem type	Assessment method	Key findings	References
1.	*Abies spectabilis, Rhododendron campanulatum, Quercus leucotrichophora*	Alpine ecosystem	Microbial biomass carbon, microbial biomass nitrogen, microbial biomass phosphorus	Among the forest, ecotone, and meadow zone, ecotone region drives most of the soil variables such as MBC, MBN, soil carbon, and nitrogen	Gupta et al. (2024)
2.	*Capillipedium parviflorum, Avena sterilis, Oplismenus compositus*	Himalayan agricultural ecosystem	Abandoned farmland—microbial biomass carbon, microbial biomass nitrogen, microbial biomass phosphorus	Revegetation following abandoned farmland changes in the soil and microbial stoichiometry, and also changes in the vegetation composition and nutrient cycling of the abandoned farmland. The microbial quotient (MBC/SOC and MBN/TN) decreased with the stand age, reflecting high nutrient uptake efficiency in young and mature stand age compared to old farmland ecosystems	Joshi et al. (2023)

(continued)

Table 3.1 (continued)

S. No.	Soil biodiversity studied	Ecosystem type	Assessment method	Key findings	References
3.	*Cedrus deodara* and *Quercus leucotrichophora*	Forest ecosystem	Rhizosphere properties—microbial biomass carbon, microbial biomass nitrogen, microbial biomass phosphorus	Plant species and stand age significantly influenced soil moisture, SOC, TN, and microbial biomass in the rhizosphere. SOC/TN ratio increased from juvenile stage to mature stage. Then it declined in the old stage. PCA results also reflected a strong coupling between juvenile stand age and MBN/TN ratio. The microbial quotient (MBC/SOC and MBN/TN) decreased with the stand age	Kumar et al. (2023a, b)
4.	*Quercus leucotrichophora*, *Pinus roxburghii*	Forest ecosystem	Soil and microbial biomass and enzyme activity	The results showed that MBC, MBC:SOC, and MBP:AP are the most important variables to evaluate rhizosphere soil quality of temperate forest	Kumar et al. (2022)
5.	*Proteobacteria*, *Phenylobacterium*, and *Candidatus solibacter*	Himalayan temperate forest	DNA extraction and metagenome (Illumina MiSeq) analysis	Disturbance level showed variation in microbial community composition. *Proteobacteria* was the most abundant phylum, and *Phenylobacterium* and *Candidatus solibacter* were mainly abundant at the genus level	Singh (2021)

(continued)

Table 3.1 (continued)

S. No.	Soil biodiversity studied	Ecosystem type	Assessment method	Key findings	References
6.	*Quercus leucotrichophora, Pinus roxburghii*	Mountain forest ecosystem	Soil nutrients and flora	Soil nutrient (carbon and nitrogen) improved with increase in tree density and basal area of the forests	Verma and Garkoti (2019)
7.	Plant invasion studies—*Lantana camara* and *Ageratina adenophora*	Forest Himalayan ecosystem	Soil amendment	The results showed that *Lantana camara* and *A. adenophora* invasion changes the soil properties and develops nutrient-rich substratum, which may promote invasiveness in chir pine forest	Kumar et al. (2021)
8.	*Alnus nepalensis*	Forest ecosystem	Microbial biomass carbon, microbial biomass nitrogen, microbial biomass phosphorus, stoichiometry	The results revealed that stoichiometry (C:N, N:P, and C:P ratios) of tree components, soil, and microbial biomass varied widely, and the presence of nitrogen-fixing *A. nepalensis* in different succession stages improved the soil and microbial biomass stoichiometry	Joshi et al. (2023)
9.	*Prosopis juliflora* and *Acacia nilotica*	Semiarid forest ecosystem in Aravalli Hills	Microbial biomass carbon, microbial biomass nitrogen, stoichiometry, stocks	Invasive *P. juliflora* altered ecosystem processes and increased the soil and microbial C and N stocks compared to native *A. nilotica*	Srivastava et al. (2023)

(continued)

Table 3.1 (continued)

S. No.	Soil biodiversity studied	Ecosystem type	Assessment method	Key findings	References
10.	Eggplant (Pusa Safed Baingan-1)	Semiarid forest ecosystem	Soil physicochemical, microbial, and nutrient analysis; soil amendment; biochars	Biochar and amended biochar combined with compost enhanced soil and plant growth parameters	Mohan et al. (2024)
11.	*Quercus leucotrichophora*, *Pinus roxburghii*	Forest ecosystem	Soil and microbial respiration	Total respiration and its components (R_{oot} respiration and microbial respiration) are highly influenced by the forest types and seasons	Kumar et al. (2022)

influences on organic matter decomposition, nutrient cycling, and plant health (Philippot et al. 2024). These interactions manifest through complex mechanisms including enzymatic degradation, nutrient solubilization, and signaling pathways. Symbiotic associations between microbes and plants, notably mycorrhizal fungi and nitrogen-fixing bacteria, facilitate nutrient acquisition, enhance stress tolerance, and modulate plant defense responses (Jacoby et al. 2017). The flowchart of soil-plant microbes has been shown in Fig. 3.2.

On the other hand, pathogenic microbes can trigger diseases, reducing plant vigor and yield, compromising plant fitness and productivity compromising plant fitness and productivity. Plant-root exudates, comprising diverse organic compounds, serve as energy sources and chemical signals, modulating microbial community composition and activity (Alberto et al. 2019). Clarifying the complexities of these interactions will hold paramount importance for optimizing agricultural productivity, fostering ecosystem resilience, and advancing sustainable land management practices.

Conclusion

It is clear from the chapter that, assessing soil microbial communities involves several methods that help us understand the interactions between soil, microbes, and plants. Traditional culture-based techniques allow us to study specific microbes but have limitations. Advanced methods like DNA sequencing and metagenomics give a complete picture of microbial diversity and functions without the need for culturing. Techniques such as stable isotope probing and metatranscriptomics help link microbial activity to nutrient cycling and plant health directly. These methods improve our understanding of how microbes support plant growth and soil health.

Table 3.2 International level literature survey on soil biodiversity studies across different ecosystems

S. No.	Soil biodiversity studied	Ecosystem type	Assessment method	Key findings	References
1	*Citrus* spp.	Agroforestry	Biochar, enzyme activity	The application of citrus peel biochars and magnesium-modified citrus peel biochar increased the soil organic carbon content, later increasing enzyme activity	Hu et al. (2023)
2	*Alpinia zerumbet*	Agriculture	Soil amendment with biochar	Biochar and compost amendments are an efficient and sustainable method for improving the metabolite contents and reducing oxidative stress in *A. zerumbet*	Zulfiqar et al. (2021)
3	Peanut yield and greenhouse gas emissions	Agriculture	Soil amendment with biochar and compost	Applications of biochar and compost have strong potential to improve organic carbon, soil water content, soil nutrient status, and peanut yield and abate greenhouse gas fluxes on tropical Ferralsol	Agegnehu et al. (2015)
4	Moso bamboo plantation	Agriculture	Soil amendment with phosphatase-solubilizing bacteria	The phosphatase-solubilizing bacterial inoculations shaped soil microbial communities and increased the moso bamboo growth by increasing soil mineral N, available phosphorus, and chlorophyll contents	Liu et al. (2023)
5	Microorganism	Agroforestry	Soil enzyme activity with fungicide application	Results showed adverse effects of copper on dehydrogenase, β-glucosidase, and phosphatase activities	Fernández-Calviño et al. (2010)

(continued)

Table 3.2 (continued)

S. No.	Soil biodiversity studied	Ecosystem type	Assessment method	Key findings	References
6	*Rhizobium meliloti* and (*Medicago sativa*)	Agriculture	Heavy metal pollution	Highest nodule number and nitrogenase activity were observed in soils which had the lowest population of *R. meliloti*	Angle and Chaney (1991)
7	Bacteria and fungi	Agriculture	Heavy metal tolerance	Bacteria of cultivated soils developed tolerance to Cu, Zn, and Cd	Rajapaksha (2011)
8	*Bacillus marisflavi* and *Trichoderma longibrachiatum*	Grassland	Heavy metal pollution remediation	Fungal strain *T. longibrachiatum* represents a viable option for Cr and Zn bioremediation, and bacterial strain *B. marisflavi* may be used in Pb bioremediation applications	Firincă et al. (2023)
9.	Maize crop	Agriculture	Soil amendment	Chemical and organic fertilizer treatment increased the relative abundance of Betaproteobacteria, Alphaproteobacteria, Gammaproteobacteria, and Gemmatimonadetes and decreased the abundance of Sordariomycetes and Agaricomycetes	Han et al. (2021)
10	*Fragaria ananassa*	Agriculture	Food and paper vermicompost	Soils treated with paper waste vermicompost contained more extractable N	Arancon et al. (2006)

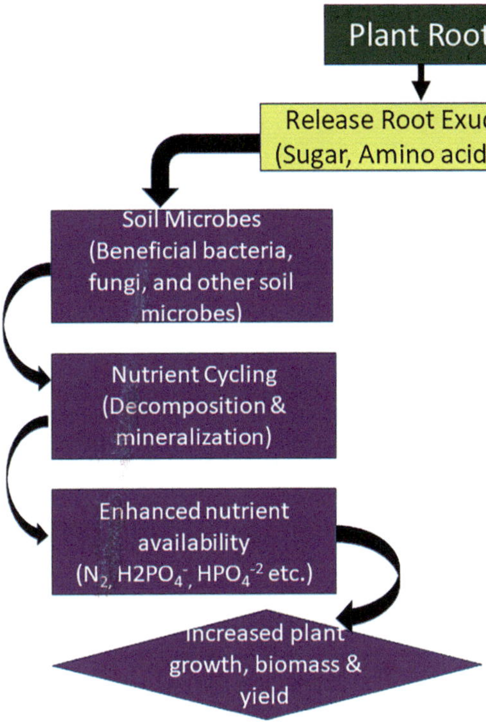

Fig. 3.2 Soil-microbe-plant interactions

As our tools and techniques improve, they will help us develop better ways to manage soil and enhance agricultural productivity sustainably.

References

Agegnehu G, Bass AM, Nelson PN, Muirhead B, Wright G, Bird MI (2015) Biochar and biochar-compost as soil amendments: effects on peanut yield, soil properties and greenhouse gas emissions in tropical North Queensland, Australia. Agric Ecosyst Environ 213:72–85

Alberto C, Christina K, Andrew M, Andreas R, Wanek W (2019) Root exudation of primary metabolites: mechanisms and their roles in plant responses to environmental stimuli. Front Plant Sci 10:Article 157. https://doi.org/10.3389/fpls.2019.00157

Angle JS, Chaney RL (1991) Heavy metal effects on soil populations and heavy metal tolerance of Rhizobium meliloti, nodulation, and growth of alfalfa. Water Air Soil Pollut 57:597–604

Arancon NQ, Edwards CA, Bierman P (2006) Influences of vermicomposts on field strawberries: Part 2. Effects on soil microbiological and chemical properties. Bioresour Technol 97(6):831–840

Artika IM, Dewi YP, Nainggolan IM, Siregar JE, Antonjaya U (2022) Real-time polymerase chain reaction: current techniques, applications, and role in COVID-19 diagnosis. Genes 13(12):Article 2387. https://doi.org/10.3390/genes13122387

Atlas RM (2010) Handbook of microbiological media, 3rd edn. CRC Press, Boca Raton, FL

Attwood TK, Blackford S, Brazas MD, Davies A, Schneider MV (2019) A global perspective on evolving bioinformatics and data science training needs. Brief Bioinform 20(2):398–404. https://doi.org/10.1093/bib/bbx100

Aytenew M (2021) Soil biodiversity as a key sponsor of regenerative agriculture. IntechOpen, London. https://doi.org/10.5772/intechopen.99716

Beattie TK, Seal DV, Tomlinson A, McFadyen AK, Grimason AM (2003) Determination of amoebicidal activities of multipurpose contact lens solutions by using a most probable number enumeration technique. J Clin Microbiol 41:2992–3000. https://doi.org/10.1128/JCM.41.7.2992-3000.2003

Bhardwaj L, Reddy B, Dubey SK (2023) Deciphering insights into rhizospheric microbial community and soil parameters under the influence of herbicides in zero-tillage tropical rice-agroecosystem. Environ Res 237:Article 117033. https://doi.org/10.1016/j.envres.2023.117033

Bhardwaj L, Kumar D, Singh U, Joshi C, Dubey S (2024) Herbicide application impacted soil microbial community composition and biochemical properties in a flooded rice field. Sci Total Environ 914:Article 169911. https://doi.org/10.1016/j.scitotenv.2024.169911

Bonnet M, Lagier JC, Raoult D, Khelaifia S (2019) Bacterial culture through selective and non-selective conditions: The evolution of culture media in clinical microbiology. New Microbes New Infect 34:Article 100622. https://doi.org/10.1016/j.nmni.2019.10062

Briones MJI (2018) The serendipitous value of soil fauna in ecosystem functioning: the unexplained explained. Front Environ Sci 6:Article 149. https://doi.org/10.3389/fenvs.2018.00149

Button M, Weber K, Nivala J et al (2016) Community-level physiological profiling of microbial communities in constructed wetlands: Effects of sample preparation. Appl Biochem Biotechnol 178:960–973. https://doi.org/10.1007/s12010-015-1921-7

Carter MR, Gregorich EG (2007) Soil sampling and methods of analysis. CRC Press, Boca Raton, FL

Chauhan P, Sharma N, Tapwal A, Kumar A, Verma GS, Meena M, Seth CS, Swapnil P (2023) Soil microbiome: diversity, benefits and interactions with plants. Sustainability 15:14643. https://doi.org/10.3390/su151914643

Cochran WG (1977) Sampling techniques, 3rd edn. Wiley, Hoboken, NJ

Coleman D (2015) Soil fauna: Occurrence, biodiversity, and roles in ecosystem function. In: Soil microbiology, ecology, and biochemistry, (4th ed.). edn. Academic Press, Cambridge, MA. https://doi.org/10.1016/B978-0-12-415955-6.00005-0

Dubey SK, Singh A, Singh RS, Upadhyay SN (2013) Changes in methanogenic population size and CH4 production potential in response to crop phenology in tropical rice field. Soil Biol Biochem 57:972–978. https://doi.org/10.1016/j.soilbio.2012.07.001

Eyler E (2013) Pouring agar plates and streaking or spreading to isolate individual colonies. In: Laboratory methods in enzymology: cell, lipid and carbohydrate. Elsevier, Amsterdam, pp 3–14. https://doi.org/10.1016/B978-0-12-420067-8.00001-5

Feng J, Liu YR, Eldridge D et al (2024) Geologically younger ecosystems are more dependent on soil biodiversity for supporting function. Nat Commun 15:Article 4141. https://doi.org/10.1038/s41467-024-48289-y

Fernández-Calviño D, Soler-Rovira P, Polo A, Díaz-Raviña M, Arias-Estévez M, Plaza C (2010) Enzyme activities in vineyard soils long-term treated with copper-based fungicides. Soil Biol Biochem 42(12):2119–2127

Fierer N, Jackson JA, Vilgalys R, Jackson RB (2005) Assessment of soil microbial community structure by use of taxon-specific quantitative PCR assays. Appl Environ Microbiol 71(7):4117–4120. https://doi.org/10.1128/AEM.71.7.4117-4120.2005

Firincă C, Zamfir LG, Constantin M, Răut I, Capră L, Popa D, Jinga ML, Baroi AM, Fierăscu RC, Corneli NO, Postolache C (2023) Microbial removal of heavy metals from contaminated environments using metal-resistant indigenous strains. J Xenobiot 14(1):51–78

Garland JL (1997) Analysis and interpretation of community-level physiological profiles in microbial ecology. FEMS Microbiol Ecol 24(4):289–300. https://doi.org/10.1111/j.1574-6941.1997.tb00446.x

Green MR, Sambrook J (2018) Touchdown polymerase chain reaction (PCR). Cold Spring Harbor Protoc 2018(1):pdb.prot095133. https://doi.org/10.1101/pdb.prot095133

Gupta R, Joshi RK, Mishra A, Kumar S, Hansda P, Garkoti SC (2024) Treeline ecotone drives the soil physical, bio-chemical and stoichiometry properties in alpine ecosystems of the western Himalaya, India. Catena 239:Article 107950

Han J, Dong Y, Zhang M (2021) Chemical fertilizer reduction with organic fertilizer effectively improves soil fertility and microbial community from newly cultivated land in the Loess Plateau of China. Appl Soil Ecol. 165:Article 103966

Hu L, Huang R, Zhou L, Qin R, He X, Deng H, Li K (2023) Effects of magnesium-modified biochar on soil organic carbon mineralization in citrus orchard. Front Microbiol. 14:Article 1109272

Jacoby R, Peukert M, Succurro A, Koprivova A, Kopriva S (2017) The role of soil microorganisms in plant mineral nutrition - current knowledge and future directions. Front Plant Sci. 8:Article 1617. https://doi.org/10.3389/fpls.2017.01617

Jezierska-Tys S, Joniec J, Mocek-Płóciniak A et al (2021) Microbial activity and community level physiological profiles (CLPP) of soil under the cultivation of spring rape with the Roundup 360 SL herbicide. J Environ Health Sci Eng 19:2013–2026. https://doi.org/10.1007/s40201-021-00753-3

Joshi RK, Garkoti SC (2023) Effect of forest chronosequence on ecological stoichiometry and nitrogen and phosphorus stocks in Alnus nepalensis forest stands, central Himalaya. Land Degrad Dev 34(12):3769–3789

Joshi RK, Garkoti SC, Gupta R, Kumar S, Mishra A, Kumar M (2023) Recovery of soil microbial biomass, stoichiometry, and herb-layer diversity with chronosequence of farmland land abandonment in the central Himalayas, India. Restor Ecol. 31(4):Article e13782

Kumar S (2012) Textbook of microbiology. JP Medical Ltd., London

Kumar S, Garkoti SC (2022) Rhizosphere influence on soil microbial biomass and enzyme activity in banj oak, chir pine and banj oak regeneration forests in the central Himalaya. Geoderma 409:Article 115626

Kumar M, Kumar S, Verma AK, Joshi RK, Garkoti SC (2021) Invasion of Lantana camara and Ageratina adenophora alters the soil physico-chemical characteristics and microbial biomass of chir pine forests in the central Himalaya, India. Catena 207:Article 105624

Kumar P, Mishra AK, Chaudhari SK, Singh R, Yadav K, Rai P, Sharma DK (2022) Conservation agriculture influences crop yield, soil carbon content and nutrient availability in the rice–wheat system of north-west India. Soil Res 60(6):624–635

Kumar S, Gupta R, Joshi RK, Garkoti SC (2023a) Rhizosphere soil properties and microbial biomass changes with the chronosequence of stand age of broadleaf banj oak and coniferous deodar forests in the central Himalaya, India. Rhizosphere 27:100761

Kumar S, Kumar M, Verma AK, Joshi RK, Hansda P, Geise A, Garkoti SC (2023b) Seasonal dynamics of soil and microbial respiration in the banj oak and chir pine forest of the central Himalaya, India. Appl Soil Ecol 182:104740

Lares-Villa F, Hernández-Peña C (2010) Concentration of Naegleria fowleri in natural waters used for recreational purposes in Sonora, Mexico. Exp Parasitol 126(1):33–36. https://doi.org/10.1016/j.exppara.2010.04.011

Lema NK, Gemeda MT, Woldesemayat AA (2023) Recent advances in metagenomic approaches, applications, and challenges. Curr Microbiol 80(4):347. https://doi.org/10.1007/s00284-023-03451-5

Lewe N, Hermans S, Lear G, Kelly LT, Thomson-Laing G, Weisbrod B, Wood SA, Keyzers RA, Deslippe JR (2021) Phospholipid fatty acid (PLFA) analysis as a tool to estimate absolute abundances from compositional 16S rRNA bacterial metabarcoding data. J Microbiol Methods 188:106271. https://doi.org/10.1016/j.mimet.2021.106271

Li B, Qin Y, Duan H, Yin W, Xia X (2011) Genome-wide characterization of new and drought stress responsive microRNAs in Populus euphratica. Journal of experimental botany, 62(11), 3765-3779

Liu Y, Nessa A, Zheng Q, Hu D, Zhang W, Zhang M (2023) Inoculations of phosphate-solubilizing bacteria alter soil microbial community and improve phosphorus bioavailability for moso bamboo (Phyllostachys edulis) growth. Appl Soil Ecol 189:104911
Lohr SL (2021) Sampling: design and analysis. Chapman and Hall/CRC, London
Luton PE, Wayne JM, Sharp RJ, Riley PW (2002) The mcrA gene as an alternative to 16S rRNA in the phylogenetic analysis of methanogen populations in landfill. Microbiology 148:3521–3530. https://doi.org/10.1099/00221287-148-11-3521
Mishra AK, Grover D, Das R, Mishra J, Biswal B (2022) Carbon sequestration potential in agricultural systems. In: Agro-biodiversity and agri-ecosystem management. Springer, Singapore, pp 87–111
Mohan D, Preetiva B, Chaubey AK, Singsit JS, Mina U, Pittman CU Jr (2024) Eggplant growth in wheat straw-, wheat straw biochar- and compost-amended soils: A field study of CO2 emission dynamics, soil physicochemical, microbial, and nutrient effects. Waste Manag Res Bull 1(4):143–157
Morris JS, Baladandayuthapani V (2017) Statistical contributions to bioinformatics: design, modeling, structure learning, and integration. Stat Modelling 17(4–5):245–289. https://doi.org/10.1177/1471082X17698255
Mullis MM, Rambo IM, Baker BJ, Reese BK (2019) Diversity, ecology, and prevalence of antimicrobials in nature. Front Microbiol 10:2518. https://doi.org/10.3389/fmicb.2019.02518
Navas-Molina JA, Peralta-Sánchez JM, González A, McMurdie PJ, Vázquez-Baeza Y, Xu Z, Ursell LK, Lauber C, Zhou H, Song SJ, Huntley J, Ackermann GL, Berg-Lyons D, Holmes S, Caporaso JG, Knight R (2013) Advancing our understanding of the human microbiome using QIIME. Methods Enzymol 531:371–444. https://doi.org/10.1016/B978-0-12-407863-5.00019-8
Roberts C, Pydipalli R, Pasam P, Anumandla SKR (2024) Species-Level Differentiation of Microbes within Multispecies Microscopic Polyculture via Complex Deep Learning Strategies. Malaysian Journal of Medical and Biological Research, 11(1), 7-20
Pérez-Cobas AE, Gomez-Valero L, Buchrieser C (2020) Metagenomic approaches in microbial ecology: an update on whole-genome and marker gene sequencing analyses. Microb Genom 6(8):000409. https://doi.org/10.1099/mgen.0.000409
Perujo N, Romaní AM, Martín-Fernández JA (2020) Microbial community-level physiological profiles: considering whole data set and integrating dynamics of colour development. Ecol Indic 117:106628. https://doi.org/10.1016/j.ecolind.2020.106628
Philippot L, Chenu C, Kappler A et al (2024) The interplay between microbial communities and soil properties. Nat Rev Microbiol 22:226–239. https://doi.org/10.1038/s41579-023-00980-5
Pinu FR, Beale DJ, Paten AM, Kouremenos K, Swarup S, Schirra HJ, Wishart D (2019) Systems biology and multi-omics integration: viewpoints from the metabolomics research community. Meta 9(4):76. https://doi.org/10.3390/metabo9040076
Power DA, Johnson JA (2009) Difco™ & BBL™ manual, 2nd edn. Franklin Lakes, NJ, Becton, Dickinson and Company
Quideau SA, McIntosh AC, Norris CE, Lloret E, Swallow MJ, Hannam K (2016) Extraction and analysis of microbial phospholipid fatty acids in soils. J Vis Exp 114:e54360. https://doi.org/10.3791/54360
Quince C, Walker A, Simpson J et al (2017) Shotgun metagenomics, from sampling to analysis. Nat Biotechnol 35:833–844. https://doi.org/10.1038/nbt.3935
Rajapaksha RMCP (2011) Heavy metal tolerance of culturable bacteria and fungi in a long-term cultivated tropical ultisol. Eur J Soil Biol 47(1):9–15. https://doi.org/10.1016/j.ejsobi.2010.10.006
Rattan L, Curtis M, Luke N, Pete S (2021) The role of soil in regulation of climate. Philos Trans R Soc B 376(20210084). https://doi.org/10.1098/rstb.2021.0084
Reyes-Sánchez L, Horn R, Costantini E, Aranda J (eds) (2022) Sustainable soil management as a key to preserve soil biodiversity and stop its degradation. International Union of Soil Sciences [IUSS]
Rowell DL (2014) Soil science: methods & applications. Routledge, London
Schoeneberger PJ, Wysocki DA, Benham EC (eds) (2012) Field book for describing and sampling soils. Government Printing

Sharpton TJ (2014) An introduction to the analysis of shotgun metagenomic data. Front Plant Sci 5:209. https://doi.org/10.3389/fpls.2014.00209

Singh S (2021) Low-to moderate-level forest disturbance effects on plant functional traits and associated soil microbial diversity in Western Himalaya. Front For Glob Change 4:710658

Srivastava S, Joshi RK, Garkoti SC (2023) Carbon and nitrogen allocation in leaf, fine root, soil, and microbial biomass in invasive Prosopis juliflora and native Acacia nilotica in semi-arid forests of the Aravalli hills. Arid Land Res Manag 37(4):535–553

Terrones-Fernandez I, Casino P, López A, Peiró S, Ríos S, Nardi-Ricart A, García-Montoya E, Asensio D, Marqués AM, Castilla R, Gamez-Montero PJ, Piqué N (2023) Improvement of the pour plate method by separate sterilization of agar and other medium components and reduction of the agar concentration. Microbiol Spectr 11(1):e0316122. https://doi.org/10.1128/spectrum.03161-22

Thompson SK (2012) Sampling, vol 755. John Wiley & Sons, Hoboken, NJ

Tibbett M, Fraser TD, Duddigan S (2020) Identifying potential threats to soil biodiversity. PeerJ. https://doi.org/10.7717/peerj.9271

Verma AK, Garkoti SC (2019) Population structure, soil characteristics and carbon stock of the regenerating banj oak forests in Almora, Central Himalaya. For Sci Technol 15(3):117–127

Wagg C, Bender SF, Widmer F, van der Heijden MG (2014) Soil biodiversity and soil community composition determine ecosystem multifunctionality. Proc Natl Acad Sci U S A. 111:5226–5270

Wagg C, Schlaeppi K, Banerjee S et al (2019) Fungal-bacterial diversity and microbiome complexity predict ecosystem functioning. Nat Commun 10:4841. https://doi.org/10.1038/s41467-019-12798-y

Wang G, Ren Y, Bai X, Su Y, Han J (2022) Contributions of beneficial microorganisms in soil remediation and quality improvement of medicinal plants. Plants (Basel) 11(23):3200. https://doi.org/10.3390/plants11233200

Webster R, Oliver MA (2007) Geostatistics for environmental scientists. John Wiley & Sons, Hoboken, NJ

Weil RR, Brady NC (2016) The nature and properties of soils, 15th edn. Pearson, Columbus, OH, p 910

Wong YP, Othman S, Lau YL, Radu S, Chee HY (2018) Loop-mediated isothermal amplification (LAMP): A versatile technique for detection of micro-organisms. J Appl Microbiol 124(3):626–643. https://doi.org/10.1111/jam.13647

Yafetto L (2022) Application of solid-state fermentation by microbial biotechnology for bioprocessing of agro-industrial wastes from 1970 to 2020: a review and bibliometric analysis. Heliyon 8(3):e09173. https://doi.org/10.1016/j.heliyon.2022.e09173

Zhang X, Kasimu A, Liang H, Wei B, Aizizi Y, Han F (2024) Mechanism analysis of vegetation phenology in an urban agglomeration in an arid zone driven by seasonal land surface temperatures. Urban Climate, 53, 101795

Zamkovaya T, Foster JS, de Crécy-Lagard V et al (2021) A network approach to elucidate and prioritize microbial dark matter in microbial communities. ISME J 15:228–244. https://doi.org/10.1038/s41396-020-00777-x

Zulfiqar F, Chen J, Younis A, Abideen Z, Naveed M, Koyro HW, Siddique KH (2021) Biochar, compost, and biochar–compost blend applications modulate growth, photosynthesis, osmolytes, and antioxidant system of medicinal plant Alpinia zerumbet. Front Plant Sci 12:707061

Defining Framework for Soil Health Assessment for Regenerative Agriculture Under Different Land Uses

4

Chetan Gyawali, Nirmal De, Sabyasachi Koley, Krisanu Golui, Charchika Khanna, and Athuman Juma Mahinda

Introduction

In recent years, there has been a growing awareness to shift toward more sustainable agricultural practices. Traditional agricultural systems, marked by heavy chemical use, large-scale monocultures, and widespread land clearing, have played a significant role in environmental degradation, soil erosion, and loss of biodiversity and climate change (Rai et al. 2023). The increasing global population, predicted to exceed 9.7 billion by the mid-century (United Nations 2019), is mounting pressures on natural resources. So, it is crucial to investigate alternative methods that can simultaneously tackle the challenges of food security and environmental sustainability.

Soil health is the cornerstone of sustainable agriculture, playing a pivotal role in ensuring the long-term viability of our food production systems (Telo da Gama 2023). Soil health, also referred to as soil quality, is defined as "the capacity of soil to function within ecosystem boundaries to sustain biological activity, maintain environmental quality, and promote plant and animal health" (Doran and Zeiss

C. Gyawali (✉) · N. De · S. Koley
Department of Soil Science and Agricultural Chemistry, Institute of Agricultural Sciences, Banaras Hindu University, Varanasi, Uttar Pradesh, India

K. Golui
Department of Soil Science, Chaudhary Charan Singh Haryana Agricultural University, Hisar, Haryana, India

C. Khanna
Department of Soil Science and Agricultural Chemistry, Sam Higginbottom University of Agriculture and Technology, Prayagraj, Uttar Pradesh, India

A. J. Mahinda
College of Agriculture and Food Technology, University of Dar es Salaam, Dar es Salaam,, Tanzania

A. K. Mishra et al. (eds.), *Key Drivers and Indicators of Soil Health Management*, https://doi.org/10.1007/978-981-97-7564-4_4

2000). Land use and management practices can induce changes in soil health and alter its functioning (Li et al. 2021). Intensive tillage-based agriculture exacerbates soil quality degradation by disrupting soil structure, hastening the oxidation of soil organic matter, and diminishing biological activity (Lal 2015a, b). These impacts pose risks to crop production and long-term soil health. To this effect, regenerative agriculture (RA) has gained increasing acknowledgment as a promising solution (Soto et al. 2021). The term "regenerative" denotes the ability to restore or renew, implying that an object or system possessing this quality has the inherent capability to re-establish itself (Christopher 2015). According to Rhodes (2017), regenerative agriculture (RA) is fundamentally centered on enhancing soil health or revitalizing severely degraded soil, which in turn synergistically improves water quality, vegetation, and land productivity. Soil health forms the bedrock of regenerative agriculture, fostering a sustainable, resilient, and productive farming system that contributes to environmental preservation and societal well-being (Tahat et al. 2020). Hence, it is crucial to possess a comprehensive understanding of soil health, its assessment methods, and effective management strategies to ensure the optimal and sustainable provision of ecosystem services by soils (Guo 2021; Mishra et al. 2022; Mishra et al. 2024). This chapter delves into the multifaceted importance of soil health assessment in regenerative agriculture, examining its impact on productivity, environmental sustainability, and resilience.

Principles and Practices of Regenerative Agriculture

RA practitioners use several methodologies that combine biological and ecological systems to increase productivity and restore landscape function. Regenerative agriculture principles (Fig. 4.1) and practices are interlinked and vital for sustaining ecosystems and enhancing agricultural lands (Mishra et al. 2024). Khangura et al. (2023) proposed five principles of RA:

1. Minimize soil disturbance: Industrial agricultural practices significantly hasten soil erosion. Consequently, the primary necessity for soil regeneration is minimizing tillage and reducing soil disturbance. Regenerative agriculture seeks to achieve this by lessening soil disruption, thereby fostering the growth of beneficial microorganisms, enhancing the water and nutrient retention capacity, sequestering carbon in the soil, and lowering greenhouse gas emissions, positioning it as a crucial strategy for combating climate change (Lal 2015a, b).
2. Keep soils covered: A layer of green crops or dried stubble can shield the soil surface from erosion caused by rain. This cover enhances the water retention, thereby improving the nutrient cycle. Regenerative agriculture strongly emphasizes maintaining and building soil armor, the protective organic matter layer on the soil surface (Blanco-Canqui and Lal 2010).
3. Keep living roots in soil year-round: A fundamental principle of regenerative agriculture is maintaining live roots in the soil throughout the year. This is accomplished by planting cover crops between main crops, which help protect

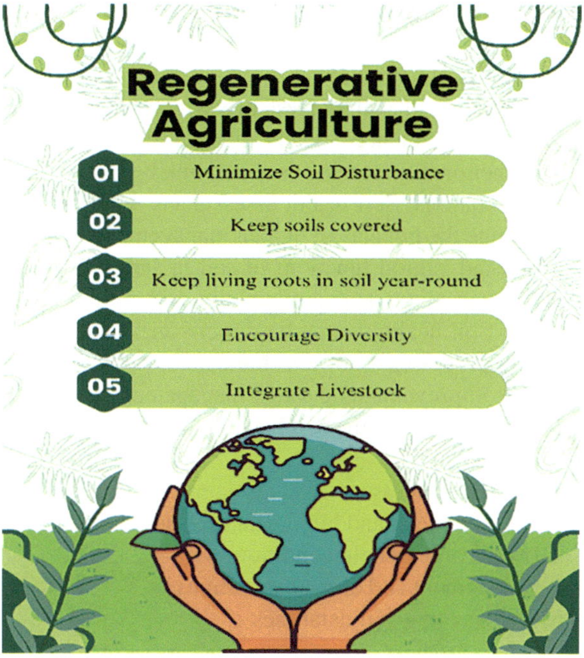

Fig. 4.1 Principle of regenerative agriculture

and enhance the soil. These cover crops are selected for their specific benefits, such as nitrogen fixation, improving soil structure, or weed suppression (Snapp and Borden 2005).

4. Encourage diversity: Crop diversity involves cultivating various crops on a farm instead of relying on a single crop. This can be achieved through cover crops, crop rotation, and intercropping. Unlike monoculture, which depletes soil nutrients and promotes erosion, crop diversity helps maintain soil nutrient balance as different plants contribute specific nutrients (Bowles et al. 2020).
5. Integrate livestock: Regenerative agriculture also integrates livestock into farming systems, raising animals like cattle, sheep, pigs, or chickens in ways that benefit the land, animals, people, and the environment. Grazing animals enhance the effects of other regenerative practices by dispersing seeds, breaking up capped soil, and facilitating gaseous exchange (Teague and Barnes 2017).

Khangura et al. (2023) also described primary land-use and farming practices for RA which include (i) no/minimum tillage, (ii) stubble retention, (iii) diverse crop rotations, (iv) multispecies cover crops, (v) intercropping, (vi) composting and use of bio-stimulants, (vii) rotational grazing, and (viii) reduction of synthetic inputs.

Framework for Soil Health Assessment

Soil health assessment is essential to regenerative agriculture because it provides information about the state of soil ecosystems, which is important for managing land sustainably. The Soil Science Society of America advocates evaluating soil health based on its functionality, serving as a vital link between human interests and environmental considerations. Soil health assessments within ecosystems offer an effective means to gauge the impacts of human management decisions, encompassing both human uses and natural functions of soil in agricultural and natural environments. This approach aligns soil productivity with soil health and quality in agricultural research. Soil health cannot be directly measured, but certain observable soil properties or processes, known as indicators, help assess the extent of soil degradation and track improvements through management practices. Doran (1996) outlined the following criteria for the soil health indicators to meet:

- Encompass ecosystem processes
- Integrate physical, chemical, and biological aspects
- Accessible and applicable to field conditions
- Sensitivity to management and climate variations
- Compatibility with existing soil databases

Changes in these indicators reflect improvements, stability, or decline in soil health due to management, land use, or conservation practices. These indicators, categorized into physical, chemical, and biological properties, form the basis for assessing soil health in agricultural contexts, although consensus is lacking on specific thresholds or values for different grades of soil health.

Soil Physical Indicators

The physical state of soil significantly impacts crop growth and environmental well-being. Properly managed soils with good aggregation maintain optimal air and water balance, fostering nutrient cycling and root growth and resisting erosion and other degradative processes. Soil physical attributes play a crucial role in productivity and are integral to soil health assessments (Cambardella et al. 1996; Wagenet and Hutson 1997) (Table 4.1).

Soil Chemical Indicators

The chemical state of soil impacts its relationship with plants by affecting nutrient holding, supply, and water movement. Assessing soil health based on chemical properties requires a defined sampling protocol, accurate chemical analysis methods, comprehension of chemical impacts on biology and minerals, contamination

Table 4.1 Review of existing soil health assessment frameworks

Soil health assessment framework	Indicators in minimum dataset (MDS)	Methodology	Applications	Strengths	Limitations	Reference
Cornell Comprehensive Assessment of Soil Health (CASH) developed by Cornell Soil Health Program	Soil texture, aggregate stability, available water capacity, soil organic matter, pH, extractable nutrients, microbial biomass, and respiration	Laboratory analysis of soil samples, with results integrated into a comprehensive soil health report	Used by researchers, agricultural extension agents, and advanced farmers to identify issues with soil health and decide strategies for land management	Comprehensive, scientifically sound, and offers practical guidance for decision-making	Requires specialized equipment and skills, involves longer processing time and high cost	Moebius-Clune et al. (2016)
Soil Management Assessment Framework (SMAF) developed by the USDA-ARS	Physical (bulk density, water infiltration), chemical (pH, nutrient levels), and biological (soil organic carbon, microbial activity)	Includes field sampling and laboratory analysis, with results evaluated through scoring functions to produce a soil health index	Used by researchers and land managers to evaluate and compare the effectiveness of various soil management practices	Holistic approach, flexible to various soil types and management objectives	Needs laboratory resources and specialized knowledge, limiting accessibility for farmers	Lisboa et al. (2019)
Haney Soil Health Test developed by Dr. R.L. Haney at the USDA-ARS	Soil respiration, water-extractable organic carbon and nitrogen, and the Haney soil health calculation	Laboratory analysis using specific procedures intended to replicate natural soil processes	Favored by regenerative and organic farmers due to its focus on soil biology and nutrient cycling	Emphasizes biological activity, easy interpretation, and encourages sustainable soil management	All aspects of soil health are not covered and requires specified lab tests	Haney et al. (2018)
Soil Health Card (SHC) launched by the Ministry of Agriculture and Farmers' Welfare, Govt. of India, in February 19, 2015	pH, electrical conductivity, organic carbon, available nutrients (N, P, K, S, Fe, Zn, Cu, Mn, and B)	Farmers gather soil samples and simple soil tests to complete the card	Tailored for farmers with basic training, aiding them in making informed choices regarding soil management techniques	Offers simplicity, affordability, and instant results	Its scope and precision are constrained, potentially failing to encompass all aspects of soil health comprehensively	Chaudhary (2018)

Table 4.2 Physical indicators of soil health

Aeration	Clay minerals	Penetration resistance
Aggregate stability	Depth of root-limiting layers	Soil structure
Bulk density	Pore connectivity	Particle size distribution
Color	Oxygen diffusion rate	Soil tilth
Hydraulic conductivity	Pore size distribution	Total porosity
Consistence	Soil strength	Temperature

Table 4.3 Chemical indicators of soil health

Cation-exchange capacity	Exchangeable sodium percent
Contaminant availability	Nutrient cycling rates
Base saturation percent	pH
Contaminant concentration	Plant nutrient availability
Contaminant mobility	Plant nutrient content
Contaminant presence	Sodium absorption ratio
Electrical conductivity	

Table 4.4 Biological indicators of soil health

Microbial biomass	Total organic C
Biomass C	Respiration/biomass
Total C and N	Microbial community fingerprinting
Total biomass	Substrate utilization
Potentially mineralizable N	Fatty acid analysis
Soil respiration	Nucleic acid analysis
Soil enzymes (phosphatase, dehydrogenase, arylsulfatase)	

detection methods, and standardized soil characterization. Certain chemical properties serve as indicators of soil health (Table 4.2).

Soil Biological Indicators

Soil health assessment encompasses various biological factors (Table 4.3). Soil ecosystems host a diverse range of organisms, from beneficial to harmful. While soil health definitions often emphasize the presence of beneficial organisms over the absence of detrimental ones, both are crucial. Biological and biochemical soil properties, particularly those involved in energy flow and nutrient cycling, are sensitive indicators of soil health, responding to even minor changes in soil conditions (Pascual et al. 2000). Soil enzyme activities are proposed as suitable indicators due to their close ties to soil biology, ease of measurement, and rapid response to changes in soil management (Table 4.4).

Soil health indicators differ based on land use, requiring customized assessment methods. For instance, in croplands, indicators such as soil organic carbon and aggregate stability are crucial (Six et al. 1999). Conversely, in pastures, factors like soil compaction and infiltration rate are more significant (Pulleman et al. 2003).

The current approaches for assessing soil health (SH) involve three main steps:

Attribute selection: Relevant chemical, biological, and physical attributes are identified to describe soil health, focusing on measurable and cost-effective indicators that reflect management practices, land use, and climate change (Kibblewhite et al. 2008).

Quantification and scoring: These attributes are measured directly, and scores are assigned to ensure robust data that accurately reflect soil health across various conditions (Fierer et al. 2012a, b).

Integration of attributes: Scored attributes are combined into an SH index, with criteria set to weight each attribute's importance, using statistical methods to ensure that the index represents soil functionality and ecosystem services accurately (Sanchez et al. 2009; McBratney et al. 2005a, b).

Methodologies for Soil Health Assessment

Soil health assessment employs various methodologies, from conventional soil nutrient analysis to advanced molecular techniques (Kibblewhite et al. 2008). While traditional tests remain prevalent, emerging methods like DNA sequencing offer deeper insights into microbial communities and their functions (Fierer et al. 2012a, b). Additionally, remote sensing and GIS significantly enhance soil health assessment, enabling spatially explicit analyses. Satellite imagery helps monitor soil moisture, vegetation cover, and land-use changes, while GIS supports spatial mapping of soil health indicators, facilitating informed decision-making (McBratney et al. 2005a, b). Artificial intelligence and sensor-based systems hold promise for revolutionizing soil health assessment (Raj et al. 2018) (Table 4.5).

Challenges and Opportunities of Using Regenerative Agriculture

Comprehensive assessment of soil health is crucial for global food production and ecosystem services, necessitating a holistic index to quantitatively evaluate soil well-being. However, challenges arise in developing such an index due to the complexity of transforming multiple indicators into a single score, including nonlinear transformations and inconsistent scoring approaches (Bünemann et al. 2018; Roper et al. 2017). Existing frameworks often prioritize soil productivity over long-term sustainability and lack of transparency in statistical methods.

To address these limitations, Rinot et al. (2018) proposed a multivariate-complex approach for soil health assessment (Fig. 4.2). This approach involves quantifying connections between soil attributes and ecosystem services, minimizing attributes using statistical techniques like autocorrelation and principal component analysis, and correlating measured soil attributes with normalized ecosystem service values using least squares models. This comprehensive methodology aims to provide a more reliable assessment of soil health and its contributions to ecosystem services.

Table 4.5 Case studies highlighting successful implementation and outcomes under different land uses

Land use	Case study	Implementation and outcomes	Reference
Croplands	Rodale Institute Farming Systems Trial	Implemented soil health assessment to compare organic vs. conventional farming systems. Organic systems showed improved soil health and crop yields	Mäder et al. (2002)
Grazing lands	Savory Institute Holistic Management	Applied holistic grazing management to restore degraded rangelands. Improved soil health indicators such as soil organic matter and water infiltration rates	Teague et al. (2016)
Orchards/ vineyards	Fetzer Vineyards Biodynamic Farming	Utilized biodynamic farming practices to enhance soil health and biodiversity in vineyards. Improved soil structure, nutrient cycling, and crop quality	Steenwerth et al. (2008)
Urban agriculture	Brooklyn Grange Rooftop Farm	Implemented composting and intensive gardening techniques on urban rooftops. Increased soil fertility, biodiversity, and food production capacity	Pouyat et al. (2015)
Rangelands	Australian Rangeland Grazing Management	Adopted planned grazing and land restoration strategies. Enhanced soil carbon sequestration and vegetation cover, reducing erosion and desertification	Eldridge et al. (2011)

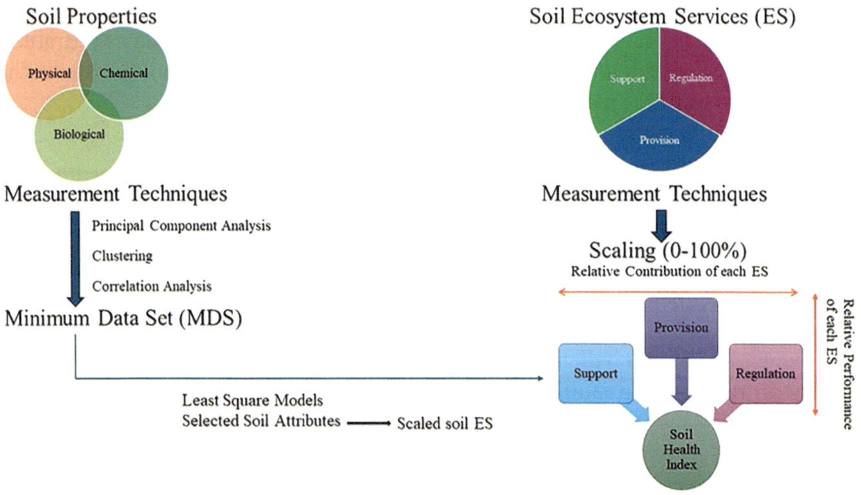

Fig. 4.2 New approach for soil health assessment to address limitation and uncertainty in existing model (adapted from Rinot et al. 2018)

Identifying research gaps and areas for future development in soil health assessment involves engaging diverse stakeholders, incorporating multi-criteria decision analysis into assessment indices, and exploring new data analysis tools like machine learning (Bampa et al. 2019; Van Joolingen et al. 2005). Collaboration across disciplines is essential to develop technologies that meet stakeholder needs and advance the recognition of soil health as a comprehensive scientific field. International standards for indicators, methodologies, and indices should be developed and endorsed to guide decisions impacting soil and its functions toward sustainability goals (Lehmann et al. 2020).

Opportunities for improving soil health management practices in diverse agricultural landscapes lie in regenerative agriculture. This approach emphasizes techniques like reduced tillage, crop rotation, and cover cropping, which promote soil health by increasing organic matter, improving soil structure, and enhancing microbial diversity (Lal 2015a, b). Incorporating livestock through rotational grazing further enhances nutrient cycling and soil fertility, supporting long-term agricultural productivity and environmental well-being (Kremen et al. 2012). Regenerative agriculture offers a holistic and sustainable approach to soil health management, addressing challenges while leveraging opportunities for agricultural sustainability.

Implementation and Adoption of Regenerative Agriculture

Implementation and adoption of regenerative agriculture require an intensive effort to engage both actors and policy makers, as well as training and capacity-building initiatives. Strategies such as targeted education campaigns, stakeholder engagement forums, and dissemination of successful case studies can effectively demonstrate the benefits of regenerative agriculture to promote adoption by stakeholders and decision makers (White et al. 2019). Training and development initiatives are crucial to equip farmers with the knowledge and skills they need. This can include hands-on workshops, technical assistance, and access to showrooms or mentoring programs. In addition, policy recommendations play a key role in supporting soil health in the context of regenerative agriculture. These recommendations may include encouraging the adoption of regenerative agricultural practices through financial support mechanisms, incorporating soil health considerations into agricultural policy, and promoting research and innovation (Kremen et al. 2012). By combining these approaches, stakeholders can work together to create an enabling environment for the widespread adoption of regenerative agriculture practices.

Conclusion

The framework for soil health assessment in regenerative agriculture, applicable across different land-use types, integrates rigorous scientific principles and practices to enhance soil health and ensure sustainable agricultural productivity. Principles of RA include minimizing soil disturbance, maintaining soil cover, sustaining living roots throughout the year, promoting crop diversity, and integrating livestock. The assessment framework rigorously evaluates soil's physical, chemical, and biological properties using both conventional methods and advanced techniques such as DNA sequencing, remote sensing, GIS, and AI for comprehensive analysis. Despite the complexity in developing a reliable soil health index, interdisciplinary research and establishment of international standards are required. Effective implementation requires promoting education, engaging stakeholders, and formulating supportive policies, which are essential for the widespread adoption of regenerative practices. This approach aims to restore soil health, enhance biodiversity, and ensure sustainable food production across diverse land-use scenarios.

References

Bampa F, O'Sullivan L, Madena K, Sandén T, Spiegel H, Henriksen CB et al (2019) Harvesting European knowledge on soil functions and land management using multi-criteria decision analysis. Soil Use Manag 35(1):6–20

Blanco-Canqui H, Lal R (2010) Soil conservation and carbon dynamics. In: Lal R (ed) Principles of soil conservation and management. Cham, Springer, pp 301–325. https://doi.org/10.1007/978-1-4020-8709-7_17

Bowles TM, Mooshammer M, Socolar Y, Calderón F, Cavigelli MA, Culman SW et al (2020) Long-term evidence shows that crop-rotation diversification increases agricultural resilience to adverse growing conditions in North America. One Earth 2(3):284–293. https://doi.org/10.1016/j.oneear.2020.02.006

Bünemann EK, Bongiorno G, Bai Z, Creamer RE, de Deyn G, de Goede R, Fleskens L, Geissen V, Kuyper TW, Mäder P, Pulleman M, Sukkel W, van Groenigen JW, Brussaard L (2018) Soil quality - a critical review. Soil Biol Biochem. https://doi.org/10.1016/j.soilbio.2018.01.030

Cambardella CA, Colvin TS, Karlen DL, Logsdon SD, Berry EC, Radke JK, et al (1996, January) Soil property contributions to yield variation patterns. In Proceedings of the Third International Conference on Precision Agriculture (pp 189–195). American Society of Agronomy, Crop Science Society of America, Soil Science Society of America

Chaudhary A (2018) The prospects and challenges of soil health card scheme. Int J Agric Sci Res 8:15–24

Christopher J (2015) Permaculture: Regenerative – not merely sustainable. Sci Prog 98:403–412. https://doi.org/10.3184/003685015X14467291596242

Doran JW (1996) Soil health and global sustainability. In: Soil Quality is in the Hands of the Land Manager. Elsevier, Amsterdam, p 45

Doran JW, Zeiss MR (2000) Soil health and sustainability: managing the biotic component of soil quality. Appl Soil Ecol 15:3–11. https://doi.org/10.1016/S0929-1393(00)00067-6

Eldridge DJ, Greene RS, Dean C (2011) Climate change impacts on soil processes in rangelands. In: Angers DA, Chantigny ML (eds) Soil health and climate change. Springer, Cham, pp 237–255

Fierer N, Lauber CL, Ramirez KS, Zaneveld J, Bradford MA, Knight R (2012a) Comparative metagenomic, phylogenetic and physiological analyses of soil microbial communities across nitrogen gradients. ISME J 6(5):1007–1017

Fierer N, Leff JW, Adams BJ, Nielsen UN, Bates ST, Lauber CL et al (2012b) Cross-biome metagenomic analyses of soil microbial communities and their functional attributes. Proc Natl Acad Sci 109(52):21390–21395

Guo M (2021) Soil health assessment and management: recent development in science and practices. Soil Syst 5:61. https://doi.org/10.3390/soil-systems5040061

Haney RL, Haney EB, Smith DR, Harmel RD, White MJ (2018) The soil health tool—theory and initial broad-scale application. Appl Soil Ecol 125:162–168. https://doi.org/10.1016/j.apsoil.2017.07.035

Khangura R, Ferris D, Wagg C, Bowyer J (2023) Regenerative agriculture—A literature review on the practices and mechanisms used to improve soil health. Sustainability 15:2338. https://doi.org/10.3390/su15032338

Kibblewhite MG, Ritz K, Swift MJ (2008) Soil health in agricultural systems. Philos Trans R Soc B 363(1492):685–701

Kremen C, Iles A, Bacon C (2012) Diversified farming systems: an agroecological, systems-based alternative to modern industrial agriculture. Ecol Soc 17(4):40

Lal R (2015a) Restoring soil quality to mitigate soil degradation. Sustainability 7:5875–5895. https://doi.org/10.3390/su7055875

Lal R (2015b) Sequestering carbon and increasing productivity by conservation agriculture. J Soil Water Conserv 70:55A–62A. https://doi.org/10.2489/jswc.70.3.55a

Lehmann J, Bossio DA, Kögel-Knabner I, Rillig MC (2020) The concept and future prospects of soil health. Nat Rev Earth Environ 1(10):544–553. https://doi.org/10.1038/s43017-020-0103-2

Li Z, Lun F, Liu M, Xiao X, Wang C, Wang L et al (2021) Rapid diagnosis of agricultural soil health: a novel soil health index based on natural soil productivity and human management. J Environ Manag 277:111402. https://doi.org/10.1016/j.jenvman.2020.111402

Lisboa IP, Cherubin MR, Satiro LS, Siqueira-Neto M, Lima RP, Gmach MR et al (2019) Applying Soil Management Assessment Framework (SMAF) on short-term sugarcane straw removal in Brazil. Ind Crop Prod 129:175–184. https://doi.org/10.1016/j.indcrop.2018.12.004

Mäder P, Fliessbach A, Dubois D, Gunst L, Fried P, Niggli U (2002) Soil fertility and biodiversity in organic farming. Science 296(5573):1694–1697

McBratney AB, Santos MLM, Minasny B (2005a) On digital soil mapping. Geoderma 117(1–2):3–52

McBratney A, Whelan B, Ancev T, Bouma J (2005b) Future directions of precision agriculture. Precis Agric 6:7–23

Mishra AK, Sinha DD, Grover D, Roohi, Mishra S, Tyagi R, Sheoran HS (2022) Regenerative agriculture as climate smart solution to improve soil health and crop productivity thereby catalysing farmers' livelihood and sustainability. Towards Sustain Nat Resour 1:295–309

Mishra AK, Bhowmick MK, Peramaiyan P, Sharma S, Singh S (2024) Regenerative agriculture practices for rice-based systems in south Asia. In: Rakshit A, Parihar M, Singh Meena V, Abhilash P, Jha PK, Sarkar D (eds) Regenerative Agriculture: Translating Science to Action, (1st edn. CRC Press, Boca Raton, FL, pp 89–94. https://doi.org/10.1201/9781003309581

Moebius-Clune BN, Moebius-Clune DJ, Gugino BK, Idowu OJ, Schindelbeck RR, Ristow AJ, van Es JE, Thies HA, Shayler MB, McBride KSMK, Wolfe DW, Abawi GS (2016) Comprehensive assessment of soil health - the Cornell framework manual, Edition, 3.2. Cornell University, Geneva, NY

Pascual JA, Garcia C, Hernandez T, Moreno JL, Ros M (2000) Soil microbial activity as a biomarker of degradation and remediation processes. Soil Biol Biochem 32(13):1877–1883

Pouyat RV, Yesilonis ID, Dombos M, Szlavecz K, Setälä H, Cilliers S, Hornung E, Kotze DJ, Yarwood S (2015) A global comparison of surface soil characteristics across five cities: a test of the urban ecosystem convergence hypothesis. Soil Sci 180(4/5):136–145

Pulleman M, Jongmans A, Marinissen J, Bouma J (2003) Effects of organic versus conventional arable farming on soil structure and organic matter dynamics in a marine loam in the Netherlands. Soil Use Manag 19(2):157–165

Rai AK, Sengupta S, Kundu M, Rani R (2023) Regenerative agriculture- A new beginning for tomorrow. Food Sci Rep 4:89–94

Raj C, Kumar V, Gupta HS (2018) Emerging technologies in soil health and plant nutrient management: status, challenges and opportunities. In: 1st International Conference on Emerging Technologies in Agriculture (ICETA-2017). Springer, Cham, pp 125–133

Rhodes CJ (2017) The imperative for regenerative agriculture. Sci Prog 100:80–129. https://doi.org/10.3184/003685017X14876775256165

Rinot O, Levy GJ, Steinberger Y, Svoray T, Eshel G (2018) Soil health assessment: a critical review of current methodologies and a proposed new approach. Sci Total Environ 648:1484–1491. https://doi.org/10.1016/j.scitotenv.2018.08.259

Roper WR, Osmond DL, Heitman JL, Wagger MG, Reberg-Horton SC (2017) Soil health indicators do not differentiate among agronomic management systems in North Carolina soils. Soil Sci Soc Am J 81(4):828–843

Sanchez PA, Ahamed S, Carré F, Hartemink AE, Hempel J, Huising J et al (2009) Digital soil map of the world. Science 325(5941):680–681

Six J, Elliott ET, Paustian K (1999) Aggregate and soil organic matter dynamics under conventional and no-tillage systems. Soil Sci Soc Am J 63(5):1350–1358

Snapp SS, Borden H (2005) Enhanced nitrogen mineralization in mowed or glyphosate treated cover crops compared to direct incorporation. Plant Soil 270:101–112. https://doi.org/10.1007/s11104-004-1310-4

Soto RL, Martínez-Mena M, Padilla MC, de Vente J (2021) Restoring soil quality of woody agroecosystems in Mediterranean drylands through regenerative agriculture. Agric Ecosyst Environ 306:107191. https://doi.org/10.1016/j.agee.2020.107191

Tahat MM, Alananbeh KM, Othman YA, Leskovar DI (2020) Soil health and sustainable agriculture. Sustainability 12:4859. https://doi.org/10.3390/su12124859

Teague R, Barnes M (2017) Grazing management that regenerates ecosystem function and grazing land livelihoods. Afr J Range Forage Sci 34:77–86. https://doi.org/10.2989/10220119.2017.1334706

Teague WR, Apfelbaum S, Lal R, Kreuter UP, Rowntree J, Davies CA, Conser R, Rasmussen M, Hatfield J, Wang T, Wang F (2016) The role of ruminants in reducing agriculture's carbon footprint in North America. J Soil Water Conserv 71(2):156–164

Telo da Gama J (2023) The role of soils in sustainability, climate change, and ecosystem services: challenges and opportunities. Ecologies 4:552–567. https://doi.org/10.3390/ecologies4030036

United Nations (2019) Department of Economic and Social Affairs, Population Dynamics. World Population Prospects 2019. Available online: https://population.un.org/wpp2019/

Van Joolingen WR, de Jong T, Lazonder AW, Savelsbergh ER, Manlove S (2005) Co-Lab: Research and development of an online learning environment for collaborative scientific discovery learning. Comput Hum Behav 21(4):671–688

Wagenet RJ, Hutson JL (1997) Soil quality and its dependence on dynamic physical processes. J Environ Qual 26(1):41–48

White K, Habib R, Hardisty DJ (2019) How to SHIFT consumer behaviors to be more sustainable: a literature review and guiding framework. J Mark 83(3):22–49

Regenerative Agriculture: Inclusive Approach for Transforming Conventional Agriculture

5

Subhrajyoti Chatterjee, Pritismita Nayak, Sumit Acharya, R. Arockia Infant Paul, and Jimmy A. Ocaña-Reyes

Introduction

Conventional farming, characterized by the excessive application of synthetic fertilizers and intensive practices, often leads to the deterioration of soil health, water pollution, erosion, and loss of habitat and biodiversity (Tsion and Steven 2019). Regenerative agriculture stands as a promising paradigm shift in the realm of farming practices, offering a sustainable alternative to conventional methods that have dominated the agricultural landscape for a long time. At its core, regenerative agriculture goes beyond mere sustainability, aiming to restore and rejuvenate the health of ecosystems while concurrently addressing the pressing environmental challenges associated with conventional farming (Ponisio et al. 2015). In the face of the environmental challenges, there is an urgent need for a transformative approach to agriculture. Regenerative agriculture presents itself as a viable solution that not only mitigates the negative impacts of conventional practices but also actively contributes to the regeneration of ecosystems (Abeliotis et al. 2013). The call for transformation is not just a response to the failures of current agricultural methods but recognition of the interconnected nature of environmental, social, and economic well-being (Rhodes 2015). By prioritizing soil health, regenerative agriculture enhances the capacity of the land to sequester carbon, retain water, and support

S. Chatterjee · P. Nayak · S. Acharya
Department of Horticulture, M S Swaminathan School of Agriculture (MSSSoA), Centurion University of Technology and Management (CUTM), Gajapati, Odisha, India

R. A. I. Paul (✉)
International Rice Research Institute South Asia Regional Center, Varanasi, Uttar Pradesh, India

J. A. Ocaña-Reyes
Estación Experimental Agraria Pucallpa, Dirección de Desarrollo Tecnológico Agrario, Instituto Nacional de Innovación Agraria (INIA), Pucallpa, Peru

A. K. Mishra et al. (eds.), *Key Drivers and Indicators of Soil Health Management*, https://doi.org/10.1007/978-981-97-7564-4_5

diverse ecosystems. It breaks away from the linear model of input-intensive farming and embraces a circular, regenerative cycle where the health of the soil is paramount (Brodd and Osanius 2002).

Historical Context

Exploring the Origins and Evolution of Regenerative Agriculture

The roots of regenerative agriculture can be traced back to early twentieth century thinkers who questioned the sustainability of industrialized farming practices. The concept of agricultural sustainability has been under discussion since the 1950s and 1960s. It gained significant attention when it became a major focus of the Brundtland Commission (1983–1987). One of the foundational concepts that laid the groundwork for regenerative agriculture was "organic farming" championed by Sir Albert Howard in the 1940s (Elshout et al. 2014). Howard emphasized the importance of maintaining soil health through natural processes and minimizing external inputs. His work laid the foundation for the organic farming movement, which served as a precursor to regenerative agriculture. In the 1970s, permaculture, a design system advocating for sustainable human habitats, emerged with Mollison and Holmgren (1981) as key figures (Smith 2014). The United Nations (UN) and other international organizations focus on several actions in rejuvenations of ecological system, particularly through regenerative approaches.

Principles of Regenerative Agriculture

The regenerative agriculture follows five major principles (Fig. 5.1). These principles help to maintain the equilibrium of ecosystem function, soil health, sustainability crop production, and reduced external application of synthetic fertilizers (Khangura et al. 2023).

1. **Soil health: Minimize soil disturbance**
   (a) **Importance**: Regenerative agriculture prioritizes soil as a living, dynamic system crucial for plant and ecosystem health.
   (b) **Practices**: Cover cropping, minimal soil disturbance, and use of organic matter enhance soil structure, fertility, and microbial diversity (Alkemade et al. 2009).
2. **Biodiversity: Maximize crop diversity**
   (a) **Importance**: Diverse ecosystems are more resilient to pests, diseases, and environmental changes.
   (b) **Practices**: Crop rotation, intercropping, and maintaining diverse plant and animal species contribute to ecological balance and reduced reliance on external inputs.
3. **Regenerative crop practices: Soil cover**

Fig. 5.1 Principles of regenerative agriculture

   (a) **Importance**: It focuses on regenerating the soil and minimizing environmental impact during crop production.
   (b) **Practices**: No-till or minimum tillage, diverse cover cropping, and agroforestry contribute to soil conservation, reduced erosion, and increased carbon sequestration (Abdalla et al. 2018).
4. **Holistic grazing: Year-round living roots**
   (a) **Importance**: Mimicking natural grazing patterns supports soil health, enhances plant growth, and improves nutrient cycling.
   (b) **Practices**: Rotational grazing where livestock are moved strategically across paddocks prevents overgrazing, allows for rest periods, and promotes a healthier balance between vegetation and animal impact (Schreefel et al. 2020).
5. **Synergistic relationships: Integrate livestock**
   (a) **Importance**: This recognizes the interconnectedness of different elements within the ecosystem.

Table 5.1 Advantage of regenerative principles and practices on environment and agricultural sustainability

Principles	Practices	Advantage of regenerative agriculture on environment and agricultural sustainability
Minimize soil disturbance	Conservation tillage, reduced tillage, zero tillage	Reduced soil erosion and soil compactness, increased soil carbon and water retention, improved microbial diversity and richness, improved soil health
Maximize crop diversity	Intercrop, mixed cropping, crop rotations, legume cropping system	Improved soil fertility, soil enrichment through biological processes, reduced dependency on synthetic fertilizers, ecological balance
Soil cover	Cover cropping, green manures	Reduced soil erosion, improved microbial diversity and soil fertility
Year-round living roots	Agroforestry, permaculture	Improved soil structure, carbon sequestration, soil and water conservation, enhanced ecosystem, resilience
Integrate livestock	Mixed farming, silvopasture, pasture cropping, rotational grazing	Restoration of soil health, improved nutrient cycling, quality food

(b) **Practices**: Integrating crops and livestock, fostering beneficial insect habitats, and utilizing companion planting are some of the practices that can be followed to enhance symbiotic relationships and reduce reliance on chemical inputs (Giller et al. 2021).

The adoption of regenerative agricultural practices resulted in more benefits in terms of environmental services and agricultural sustainability (Table 5.1). It offers exclusive approaches to reduce greenhouse gas emissions and soil erosion while improving crop resilience, soil health, and carbon sequestration. In rice-based agricultural systems, regenerative farming techniques can increase yields, enhance grain quality, and minimize greenhouse gas emissions in South Asia (Mishra et al. 2024).

Environmental Impact: A Comparative Analysis of Conventional and Regenerative Agriculture

Comparing Conventional and Regenerative Agriculture

Conventional Agriculture

Conventional agricultural practices, which involve large-scale monoculture, intensive chemical inputs, and extensive land use, have been associated with significant environmental impacts. One of the primary concerns is soil degradation caused by excessive use of synthetic fertilizers and pesticides, resulting in nutrient imbalances and a decline in soil biodiversity.

Regenerative Agriculture

On the other hand, regenerative agriculture places ecological health at the forefront, with the goal of establishing a harmonious connection between farming methods and the environment. This approach focuses on key principles like soil health, biodiversity, and holistic management, aiming to reduce environmental impact while simultaneously improving ecosystem services (Codur and Watson 2018).

Positive Outcomes from Regenerative Agriculture

1. **Soil health:**
   (a) Conventional agriculture: Soil degradation is a common issue, with reduced organic matter, nutrient depletion, and increased susceptibility to erosion.
   (b) Regenerative agriculture: Practices like cover cropping, minimal soil disturbance, and organic inputs promote soil structure, microbial diversity, and carbon sequestration. Studies indicate that regenerative methods lead to improved soil health over time (Teague 2018; Bayer et al. 2015). Adopting regenerative practices as nature-based solution aids in achieving the sustainable development goal and safeguarding the soil from further degradation (Grover et al. 2024).
2. **Biodiversity:**
   (a) Conventional agriculture: Monoculture systems can lead to habitat loss, reduced wildlife diversity, and increased vulnerability to pests.
   (b) Regenerative agriculture: Diverse crop rotations, intercropping, and natural habitat preservation enhance biodiversity on and around farms (Aguilera et al. 2013).
3. **Water management:**
   (a) Conventional agriculture: Excessive irrigation and poor water retention can lead to water pollution, depletion of aquifers, and increased vulnerability to drought.
   (b) Regenerative agriculture: Water conservation practices, such as rainwater harvesting, contour plowing, and improved soil structure, contribute to more sustainable water management.
4. **Carbon sequestration:**
   (a) Conventional agriculture: Intensive plowing and chemical inputs can lead to carbon loss from the soil contributing to greenhouse gas emissions.
   (b) Regenerative agriculture: Emphasis on cover cropping, no-till farming, and agroforestry enhances carbon sequestration in soils.
5. **Pesticide use:**
   (a) Conventional agriculture: Reliance on synthetic pesticides can result in negative impacts on nontarget species, including pollinators, and contribute to the development of pesticide-resistant pests.
   (b) Regenerative agriculture: Integrated pest management, natural predators, and diverse ecosystems reduce the need for synthetic pesticides.
6. **Fertilizer use:**
   (a) Conventional agriculture: Indiscriminate use of synthetic fertilizers results in soil pollution and negatively impacts soil health.

(b) Regenerative agriculture: Integrated nutrient management and legume intercropping reduce the need for synthetic fertilizers (Mishra et al. 2022).

While both conventional and regenerative agriculture systems aim to meet global food demands, regenerative agriculture stands out as a more sustainable and environmentally friendly approach. Studies consistently highlight the positive outcomes of regenerative practices, including improved soil health, enhanced biodiversity and reduced environmental pollution.

Environmental Impact Comparison

Conventional vs. Regenerative Agriculture

Conventional agriculture often relies on intensive chemical inputs, contributing to soil degradation, water pollution, and biodiversity loss. In contrast, regenerative agriculture emphasizes sustainable practices like cover cropping and crop rotation to enhance soil health, water retention, and carbon sequestration. This approach aims to mitigate environmental impact, fostering resilient ecosystems. Transitioning to regenerative methods can promote long-term sustainability and biodiversity conservation, reducing the ecological footprint associated with conventional farming practices (Mishra et al. 2022; Mishra et al. 2024).

Social Implications of Adopting Regenerative Agriculture

Community Engagement

Regenerative agriculture fosters community engagement by emphasizing local collaboration and shared knowledge. Farmers adopting regenerative practices often participate in cooperative initiatives, knowledge-sharing networks, and community-supported agriculture (CSA) programs. This collaborative approach enhances social capital, strengthens community ties, and empowers local resilience (Briske et al. 2008).

Food Security

Regenerative agriculture contributes to food security by promoting diversified crops, reducing reliance on external inputs, and enhancing soil fertility. Moreover, community-based agriculture models enhance access to fresh, locally produced food, addressing issues of food deserts and promoting equitable distribution (Graves et al. 2004).

Challenges and Opportunities

Challenges include initial resistance to change, knowledge gaps, and need for supportive policies. However, the social benefits, such as strengthened community bonds and improved local food systems, present opportunities for fostering sustainable agricultural practices that positively impact both farmers and the broader community.

Barriers in Adoption of Regenerative Agriculture

Challenges Faced by Farmers

- Transition costs: Initial investment in regenerative practices, such as cover cropping, may pose financial challenges for farmers.
- Resistance to change: Established farming practices and skepticism towards new methods can impede the willingness of some farmers to adopt regenerative approaches.

Policy Barriers

- Lack of incentives: Limited government incentives or subsidies for regenerative practices may discourage farmers from making the initial investment required.
- Policy misalignment: Existing agricultural policies and regulations may favor conventional methods, making it difficult for farmers to transition without policy support.

Market Barriers

- Limited market access: Insufficient market demand for regenerative products may discourage farmers from adopting these practices, as they face challenges in selling their produce.

Future Directions in Regenerative Agriculture

Emerging Trends and Technologies

- The future scope of regenerative agriculture is intricately linked with cutting-edge technologies, including drones and artificial intelligence (AI). Drones, equipped with advanced sensors and imaging technology, revolutionize farming by providing real-time data on crop health, soil conditions, and irrigation needs. They enable farmers to monitor vast areas efficiently, facilitating precision agriculture. Drones contribute to regenerative practices by optimizing resource use, minimizing chemical inputs, and promoting sustainable land management.
- Artificial intelligence plays a pivotal role in data analysis and decision-making for regenerative agriculture. AI algorithms process large datasets from various sources, including drones and sensors, to offer insights into crop patterns, pest control, and optimal planting practices. Machine learning models can predict crop yields, identify potential issues, and recommend personalized strategies for soil improvement. This enhances the efficiency of regenerative practices, ensuring a more targeted and environmentally friendly approach.

- Together, drones and AI empower farmers to implement regenerative agriculture with greater precision and effectiveness. By leveraging these technologies, the future of regenerative agriculture holds the promise of increased sustainability, improved resource management, and a positive environmental impact on global food production (Lesur-Dumoulin et al. 2017).

Conclusion

The exploration of regenerative agriculture reveals a paradigm shift that has transformative potential for our global food systems. Regenerative agriculture is a holistic approach that prioritizes soil health, biodiversity, and water management. Studies consistently demonstrate its positive environmental impact, including improved soil structure, enhanced biodiversity, and carbon sequestration. Economically, regenerative practices demonstrate viability through long-term soil health, reduced input costs, and diversified income streams. While challenges exist, such as initial transition costs and knowledge gaps, the potential for increased yields and reduced environmental impact offers economic incentives for farmers. On the social front, regenerative agriculture fosters community engagement and contributes to food security. Emerging trends and technologies, such as precision agriculture and agroecological principles, further enhance the scalability and applicability of regenerative practices. Several research projects are being conducted worldwide on regenerative agriculture to lead us towards a more sustainable, resilient, and ecologically harmonious agricultural future. It is a call to action for farmers, policymakers, and communities to collectively invest in the regenerative potential that lies within our soils, fostering a regenerative revolution for generations to come.

References

Abdalla M, Hastings A, Chadwick DR, Jones DL, Evans CD, Jones MB, Rees RM, Smith P (2018) Critical review of the impacts of grazing intensity on soil organic carbon storage and other soil quality indicators in extensively managed grasslands. Agric Ecosyst Environ 253:62–81

Abeliotis K, Detsis V, Pappia C (2013) Life cycle assessment of bean production in the Prespa National Park, Greece. J Cleaner Prod 41:89–96

Aguilera E, Lassaletta L, Gattinger A, Gimenoe BS (2013) Managing soil carbon for climate change mitigation and adaptation in Mediterranean cropping systems: a metaanalysis. Agric Ecosyst Environ 168:25–36

Alkemade R, van Oorschot M, Miles L, Nellemann C, Bakkenes M, ten Brink B (2009) GLOBIO3: A framework to investigate options for reducing global terrestrial biodiversity loss. Ecosystems 12:374–390

Bayer C, Gomes J, Zanatta JA, Vieira FCB, de Cássia PM, Dieckow J, Six J (2015) Soil nitrous oxide emissions as affected by long-term tillage, cropping systems and nitrogen fertilization in Southern Brazil. Soil Tillage Res 146:213–222

Briske D, Derner J, Brown J, Fuhlendorf S, Teague R, Gillen B, Ash A, Havstad K, Willms W (2008) Benefits of rotational grazing on rangelands: an evaluation of the experimental evidence. Rangeland Ecol Manag 61:3–17

Brodd P, Osanius M (2002) Erosion and land use at plot and catchment levels - a case study in 1 northern Vietnam. Minor Field Studies International Office, Swedish University of Agricultural 2 Sciences, Uppsala, Sweden, p 62

Codur A, Watson J (2018) Climate smart or regenerative agriculture Defining climate policies based on soil health. Global Development and Environment Institute, Tufts University. Climate Policy Brief No. 9. April 2018. http://www.ase.tufts.edu/gdae/Pubs/climate/ClimatePolicyBrief9.pdf

Elshout PMF, van Zelm R, Karuppiah R, Laurenzi IJ, Huijbregts MAJ (2014) A spatially explicit data-driven approach to assess the effect of agricultural land occupation on species groups. Int J Life Cycle Assess 19:758–769

Giller KE, Hijbeek R, Andersson JA, Sumberg J (2021) Regenerative agriculture: an agronomic perspective. Outlook Agric 50:13–25

Graves A, Matthews R, Waldie K (2004) Low external input technologies for livelihood improvement in subsistence agriculture. Adv Agron 82:473–555

Grover D, Mishra AK, Rani P, Kalonia N, Chaudhary A, Sharma S (2024) Soil management in sustainable agriculture: principles and techniques. In: Technological approaches for climate smart agriculture. Springer, Cham, pp 41–77

Khangura R, Ferris D, Wagg C, Bowyer J (2023) Regenerative agriculture - A literature review on the practices and mechanisms used to improve soil health. Sustainability 15(3):2338

Lesur-Dumoulin C, Malézieux E, Ben-Ari T, Langlais C, Makowski D (2017) Lower average yields but similar yield variability in organic versus conventional horticulture. A meta-analysis. Agron Sustain Dev 37:45

Mishra AK, Sinha DD, Grover D, Roohi, Mishra S, Tyagi R, Sheoran HS, Sharma S (2022) Regenerative agriculture as climate smart solution to improve soil health and crop productivity thereby catalysing farmers' livelihood and sustainability. In: Towards sustainable natural resources: monitoring and managing ecosystem biodiversity. Springer, Cham, pp 295–309

Mishra AK, Bhowmick MK, Peramaiyan P, Sharma S, Singh S (2024) Regenerative agriculture practices for rice-based systems in South Asia. In: Regenerative agriculture. CRC Press, Boca Raton, FL, pp 210–226

Mollison B, Holmgren D (1981) Permaculture one: a perennial agriculture for human settlements. International Tree Corps Institute, USA

Ponisio LC, M'Gonigle LK, Mace KC, Palomino J, de Valpine P, Kremen C (2015) Diversification practices reduce organic to conventional yield gap. Proc R Soc B 282:1396

Rhodes CJ (2015) Permaculture: Regenerative – not merely sustainable. https://www.resilience.org/stories/2015-11-02/permaculture-regenerative-not-merelysustainable/

Schreefel L, Schulte RP, De Boer IJM, Schrijver AP, Van Zanten HHE (2020) Regenerative agriculture–the soil is the base. Glob Food Sec 26:100404

Smith P, Clark H, Dong H, Elsiddig EA, Haberl H, Harper R, House J, Jafari M, et al (2014) Agriculture, forestry and other land use (AFOLU). In: Climate Change 2014: Mitigation of Climate Change. IPCC Working Group III Contribution to AR5. Cambridge University Press, Cambridge. http://pure.iiasa.ac.at/id/eprint/11115/

Teague WR (2018) Forages and Pastures Symposium: Cover crops in livestock production: Whole System Approach: Managing grazing to restore soil health and farm livelihoods. J Anim Sci 96:1519–1530

Tsion K, Steven W (2019) An overview use and impact of organic and synthetic farm inputs in developed and developing countries. Afr J Food Agric Nutr Dev 19:14517–14540

Digital Innovations for Agro-Advisory Services for Soil and Land Management

6

Sarita Mishra, H. S. Roopa, Udita Ghosh Sarkar, and Amber Rai

Introduction

Background and Importance of Digital Innovations in Agriculture

Agriculture faces numerous challenges on both global and local scales. The demand for food and other ecosystem services provided by the agricultural sector is increasing. According to the FPO, the world population will increase and reach around 9.1 billion people by 2050, and the food production should increase to about 70% keeping 2007 as base. The food production in the developing countries should increase twofold. Simultaneously, agriculture contends with accumulating economic, environmental, and institutional shocks, such as extreme weather events, market volatility, pandemics, and conflicts. These factors diminish the resilience of the agricultural and food sectors (Finger 2023). Moreover, the massive environmental footprints of the agri-food system necessitate urgent reduction. This involves addressing issues like agroecological resource overuse, environmental pollution from inorganic fertilizers and pesticides, greenhouse gas emissions, and biodiversity loss (Finger 2023). While foundational to farming communities for centuries, traditional agricultural practices often suffer from inefficiency, lack of precision, and susceptibility to environmental changes.

In recent years, digital technologies have revolutionized many industries, including agriculture. These innovations can potentially address the challenges faced by traditional farming and significantly improve agricultural productivity and sustainability. Digital innovations in agriculture encompass a wide range of technologies,

S. Mishra · H. S. Roopa (✉) · A. Rai
AgriPath, Grameen Foundation India Private Limited (GFI), Gurgaon, Haryana, India
e-mail: rsuresh@grameenfoundation.in

U. G. Sarkar
Client Insights for Impact, Grameen Foundation India Private Limited (GFI), Gurgaon, Haryana, India

A. K. Mishra et al. (eds.), *Key Drivers and Indicators of Soil Health Management*, https://doi.org/10.1007/978-981-97-7564-4_6

including information and communication technologies (ICTs), mobile applications, precision farming tools, data analytics, etc. These technologies enable farmers to make more informed decisions, optimize resource use, and increase crop yields. Integrating digital tools in agriculture is not just about adopting new technologies, but it also involves a fundamental shift in how agricultural information is generated, disseminated, and utilized.

Digitization has the potential to bring remarkable changes to the rural farming sector. To feed the rising population of developing countries like India, farming needs to become digital to make it more efficient, holistic, inclusive, and sustainable (Roy and Saha 2022).

Scope and Objectives of the Chapter

This chapter aims to explore the various digital innovations that have transformed agro-advisory services for soil and land management. The objectives are to:

- Provide an overview of agro-advisory services and their importance in modern agriculture
- Compare traditional and digital agro-advisory services, highlighting their respective advantages and challenges
- Discuss the evolution of digital technologies in agriculture and their role in enhancing agro-advisory services
- Examine different delivery models for digital agro-advisory services, including government-led initiatives, private-sector innovations, and community-based models
- Explore the role of mobile applications in various aspects of agriculture, such as seed selection, weed management, nutrient management, crop management, and harvest management
- Present case studies of successful digital agro-advisory services, assessing their impact and identifying lessons for scaling up and replication
- Analyze the challenges and limitations of digital agro-advisory services, including technological, socio-economic, and policy-related barriers
- Highlight future trends and opportunities in digital agro-advisory services, focusing on emerging technologies and their integration with sustainable farming practices

Overview of Agro-Advisory Services

Definition and Importance of Agro-Advisory Services

Agro-advisory services are suggestions for farmers, including agronomical, pest, disease, water, input management, and market-related information that facilitate decisions responding to farm changes. The primary aims are better productivity, increased profitability, and ensuring enhanced sustainability. Some of the recent agro-advisory services are theme based, which will be providing tailored advisory services that will help farmers in optimum use of the resources, risk management, and adoption of sustainable best practices. Agro-advisory services provide a promising tool for betterment of the farming community worldwide.

Traditional vs. Digital Agro-Advisory Services

Traditional agro-advisory services have long served as a crucial bridge connecting farmers with evidence-based research, education, and new technologies. These services, delivered through agricultural extension, facilitate informed decision-making, leading to improved productivity, income, and sustainable agricultural practices (Ghimire et al. 2021). Traditional methods include face-to-face interactions, printed materials, radio broadcasts, and television programs. However, these approaches face significant challenges, such as limited accessibility, infrastructural constraints, limited capacity, and management issues (Birner and Anderson 2007).

In contrast, digital agro-advisory services leverage advanced information and communication technologies (ICTs) to enhance the efficiency and reach of agricultural extension. New digital tools, such as the Internet, videos, text messages, and mobile applications, have become integral means of communication between farmers and other stakeholders (Davis and Alex 2020). These digital platforms enable the rapid dissemination of tailored information, improve accessibility for farmers in remote areas, and foster stronger connections among all actors in the agricultural value chain.

Challenges in Traditional Agro-Advisory Services

Traditional agro-advisory services face several challenges that hinder their effectiveness and reach:

- **Limited accessibility and infrastructural limitations:** Many farmers, particularly those in remote or underserved areas, lack access to extension services due to inadequate infrastructure, transportation issues, and insufficient extension personnel.
- **Capacity and management problems:** Extension services often suffer from limited capacity, inadequate training for extension workers, and poor management, leading to inconsistent and sometimes unreliable advisory services (Birner and Anderson 2007).
- **Resource constraints:** Extension services require substantial resources for training, transportation, and materials, which can be a significant burden, particularly in developing countries with limited budgets (Fuglie et al. 2020).

- **Communication barriers:** Traditional methods of communication, such as face-to-face interactions and printed materials, may not effectively reach all farmers, especially those who are illiterate or speak different languages.
- **Slow information dissemination:** The time taken to collect, process, and disseminate information through traditional channels can result in outdated advice, which may not be timely enough to address immediate agricultural challenges.
- **Variability in quality:** The quality and accuracy of advice can vary significantly depending on the knowledge and expertise of individual extension workers, leading to disparities in service delivery.
- The development of advanced ICT tools offers potential solutions to many of these challenges. Digital agro-advisory services can provide real-time, customized information to a broader audience, reduce costs, and improve the overall efficiency and effectiveness of agricultural extension. However, the adoption of digital tools also requires addressing issues such as digital literacy, Internet connectivity, and integration of traditional and digital advisory methods to ensure that all farmers benefit from these advancements (Davis and Alex 2020).

Digital Innovations in Agro-Advisory Services

Evolution of Digital Technologies in Agriculture

The evolution of digital technologies in agriculture has progressed significantly from early information and communication technology (ICT) tools like radio and television to advanced innovations such as the Internet, mobile applications, and precision farming technologies. These advancements have revolutionized agricultural practices, enhancing productivity, efficiency, and sustainability by providing farmers with real-time, data-driven insights and tools (Davis and Alex 2020).

Role of ICT in Agro-Advisory

Information and communication technologies (ICTs) play a crucial role in modern agro-advisory services by facilitating the rapid dissemination of information and connecting farmers with experts and resources. ICT tools, such as mobile applications, Internet platforms, and SMS services, provide real-time data on weather, market trends, pest management, and crop health. These technologies enhance farmers' decision-making capabilities, improve resource efficiency, and increase overall agricultural productivity (Davis and Alex 2020).

Benefits of Digital Agro-Advisory Services

Digital agro-advisory services offer a wide range of benefits, including increased timely accessibility of relevant information, improved and accurate range of agricultural practices, and enhanced digital connectivity among farmers and stakeholders, thereby resulting in resource-use efficiency, reducing better cost management, and mitigating risks associated with climate change and market volatility. Additionally, digital platforms promote sustainable farming practices and boost productivity, leading to higher incomes and better livelihoods for farmers (Fuglie et al. 2020).

Delivery Models for Agro-Advisory Services

Government-Led Initiatives

Government-led initiatives in agro-advisory services are designed to support farmers by providing essential agricultural information, training, and resources through organized and structured programs. These initiatives leverage various platforms, including traditional methods and digital technologies, to reach a broad audience and ensure that farmers receive timely and relevant information. Complementing these government efforts, private players in agro-digital information have emerged as significant contributors to the agricultural sector. Companies such as agri-tech and large agribusiness firms are developing innovative digital solutions, including mobile apps, data analytics platforms, and IoT-based tools, to provide farmers with customized recommendations. These private initiatives often focus on precision agriculture, market intelligence, and supply-chain optimization, offering farmers advanced technological resources to enhance their productivity and profitability. By collaborating with government programs, private-sector initiatives help create a more robust and comprehensive support system for farmers, driving the modernization and sustainability of agriculture.

Case Studies and Examples

- **Reuters Market Light (RML) SMS services in Uttarakhand**: The study conducted on 64 farmers in Bageshwar district of Uttarakhand, using RML's SMS service for agro-advisories, highlights the effectiveness of government-supported digital initiatives. The study aimed to analyze the socio-economic characteristics of the users, their sources of agricultural information, and their willingness to pay for SMS services. It was found that 70.31% of the beneficiaries were willing to pay for accessing RML's SMS, with a majority willing to pay Rs. 201–300 per annum. The most preferred advisories were on pest and disease management, followed by improved vegetable production technology. The willingness to pay was significantly influenced by age, education, and training experience.
- **CABI's Direct2Farm (D2F) mobile services**: A cross-sectional survey of farmers registered on CABI's Direct2Farm (D2F) user database was conducted using telephone interviews, household surveys, and focus group discussions across six states in India. The results indicated that while mobile services were ranked fifth as a source of information, they reached a large number of farmers quickly compared to traditional methods. The services significantly influenced farmers to adopt new agricultural practices ($p < 0.01$) (Kansiime et al. 2019) (Table 6.1).

Table 6.1 Some of the mobile-based agriculture initiatives from India

S. No.	Agriculture initiative	Details
1	IFFCO Kisan Sanchar Limited (IKSL)	Audio-based advisories and alerts on agricultural and allied sectors, weather, and markets
2	Reuters Market Light (RML) paid service	Text-based advisories on agricultural and allied sectors, weather, and markets
3	Commonwealth Agricultural Bureaux International (CABI) Direct to Farm (D2F)	Text-based advisories on agricultural practices, particularly pest and disease management
4	MS Swaminathan Research Foundation (MSSRF)	Text and audio advisories on agriculture, fisheries, weather, and market
5	Indian Meteorological Department (IMD)	Text-based Agromet advisory services to farmers
6	e-Agriculture	Developed by Tamil Nadu Agricultural University. Text and audio advisories and alerts on agriculture and allied sectors, weather, and markets
7	e-Sagu	Text-, audio-, and image-based advisories on agriculture and horticultural crops, weather, and markets
8	Behtar Zindagi Developed by Idea Cellular	Developed by Idea Cellular. Advisories on agriculture and allied sectors, weather, and markets
9	mKisan	Consortium of International Labour Research Institute (ILRI), Digital Green, Handygo, and CABI. Advisor mKisan Consortium of International Labour Research Institute (ILRI), Digital Green, Handygo, and CABI. Advisories on agriculture and allied sectors, weather, and markets
10	Mkrishi	Developed by Tata Consultancy Services (TCS). Information on agriculture and allied sectors, weather, and markets
11	Avaaj Otalo	Developed by IBM India Research Laboratory and the Development Support Centre, an NGO in Gujarat. Advisories on agriculture and allied sectors, weather, and markets
12	voice Krishi Vigyan Kendra (vKVK)	Developed by the Indian Institute of Technology (IIT), Kanpur. Used by Krishi Vigyan Kendras (KVKs). Audio-based advisories on agriculture and allied sectors, weather, and markets. Facility for two-way communication between farmers and experts
13	Advisories by ICAR institutes and state agricultural universities in collaboration with national informatics centers and private firms	Text-based advisories and alerts on agricultural and allied sectors, weather, and markets. Interactive Information Dissemination System (IIDS) developed by Media Lab Asia. Text-, audio-, and image-based advisories on selected crops, weather, and markets
14	Farmer SMS portal	Developed by the Ministry of Agriculture, Government of India. Used by KVKs, ICAR institutes, and state agricultural departments. Text-based advisories and alerts on agricultural and alert on allied sectors, weather, and markets

Private-Sector Innovations

Private-sector innovations in agro-advisory services leverage advanced technologies and business models to provide targeted, efficient, and scalable solutions to farmers. These innovations often address the limitations of traditional agricultural extension services by offering real-time data, precision farming tools, and personalized recommendations.

Agribusiness Models

Agribusiness models are the structural approach in executing agricultural operations starting from land preparation to production, postharvest, processing, and market linkage of agricultural products and services. In the recent past, many digital tools and platforms have been in use for enhancement of the services they provide to farmers. Such tools and platforms are individually developed and maintained by government or semi-government or private sector or in combinations.

- **Bayer Digital Farming Solutions**: Bayer offers a range of digital tools under its Climate FieldView platform, which integrates data from various sources to provide farmers with insights on field health, crop performance, and weather patterns. This helps farmers make informed decisions on planting, fertilization, and pest management, thereby increasing yields and reducing costs (Bayer 2020).
- **John Deere precision agriculture**: John Deere has developed a suite of precision agriculture tools, including advanced machinery equipped with GPS technology, sensors, and data analytics software. These tools enable farmers to optimize planting, irrigation, and harvesting processes, leading to higher efficiency and productivity. John Deere's solutions are widely used in both developed and developing countries, showcasing the scalability of their agribusiness model (John Deere 2021).

Start-Ups and Agri-Tech Companies

Start-ups and agri-tech companies are at the forefront of agricultural innovation, developing cutting-edge technologies and platforms to solve specific challenges faced by farmers. These companies often focus on niche areas and leverage modern technologies such as artificial intelligence (AI), machine learning (ML), and the Internet of Things (IoT).

- **CropIn Technology**: CropIn is an agri-tech start-up offering smart farming solutions through its SaaS-based platform to digitize farm management, enhance traceability, and improve crop health monitoring. The platform provides 360-degree farm management solutions, enabling farmers to monitor crop health and manage resources efficiently. CropIn's technology uses satellite imagery to provide information to its stakeholders, gathers and disseminates required weather data, and uses expert's predictive analytics to provide actionable insights to farmers, thereby helping farmers in optimizing operations resulting in better productivity and enhanced profitability (CropIn 2020).

- **FarmLogs:** FarmLogs is a start-up that provides digital farm management tools designed to help farmers track field activities, monitor crop health, and manage finances. The platform uses data analytics and satellite imagery to give farmers real-time insights into soil conditions, weather forecasts, and crop performance. This information helps farmers make data-driven decisions that enhance yield and profitability (FarmLogs 2021).

These private-sector innovations highlight the transformative potential of digital technologies in agriculture. By providing farmers with access to advanced tools and data-driven insights, these companies help to create a more efficient, sustainable, and profitable agricultural sector.

Public-Private Partnerships (PPPs)

Public-private partnerships (PPPs) in agro-advisory services involve collaboration between government departments and private companies to improve agricultural extension services. These partnerships aim to increase the reach, efficiency, and impact of advisory services for farmers, fostering innovation and resource sharing to support agricultural development and productivity better.

PPPs can take a variety of forms. They are not limited to bilateral collaboration between a government agency and a private corporation. PPP for sustainable agricultural development can also include, for example, multi-partner structures that bring together private companies with entities such as nongovernmental organizations (NGOs), university research institutes, and foundations. These structures have sometimes been termed "hybrid value chains" that create shared value. Broadly, PPPs include both these forms and the many other possible for-profit/not-for-profit combinations. Whatever form they take, successful PPPs have a number of features in common. The rationale for their creation is always the same: to achieve more through partnership than any of the parties could do on their own (Ganeshkumar and Dilipkumar 2021).

Synergies and Challenges

Synergies

- Resource sharing: PPPs allow for combining of resources, which includes financial investments, technical expertise, and infrastructure. This combined effort can lead to the development and deployment of more comprehensive effective and efficient agro-advisory services.
- Innovation and efficiency: Private-sector involvement often brings innovation and efficiency, utilizing advanced technologies and market-driven approaches with the problem-solving approach to tackle agricultural challenges. This complements the public sector's reach and regulatory support.
- Extended reach: Government networks and policies across the country can facilitate wider dissemination of private-sector innovations, ensuring that more farm-

ers benefit from advanced agro-advisory services by working meticulously on the need-based solutions.
- Capacity building: Training and capacity-building initiatives can be more robust when both sectors collaborate and work together, thereby enhancing the skills and knowledge of farmers and extension workers.

Challenges
- Coordination issues: Effective coordination between public and private entities can be challenging due to differing objectives, priorities, and operational cultures.
- Sustainability: Ensuring the long-term sustainability of PPP initiatives can be difficult, particularly if initial funding and support wane over time.
- Regulatory and policy barriers: Navigating regulatory and policy frameworks can be complex and time-consuming, potentially delaying the implementation of innovative solutions.
- Equity concerns: Ensuring that the benefits of PPPs reach smallholder and marginalized farmers can be challenging, as private companies might prioritize profitable ventures over social equity.

Community-Based Models

Community-based models for agro-advisory services emphasize local engagement and the active participation of farmers in understanding, adopting, and decision-making processes. These models foster a sense of ownership and empowerment among farmers, enhancing the adoption of sustainable agricultural practices.

Farmer Cooperatives and Associations
Farmer cooperatives and associations play a crucial role in community-based agro-advisory models. These organizations facilitate collective action and mutual support among farmers, enabling them to share resources, knowledge, and best practices.

- Kenya's dairy cooperatives: These cooperatives help smallholder dairy farmers improve their production and marketing capabilities. By pooling resources, farmers can access better inputs, training, and market opportunities, leading to increased productivity and incomes.
- India's Amul cooperative: The Amul cooperative is a successful example of a farmer-led organization that has transformed the dairy industry in India. It provides members with access to veterinary services, feed, and markets, ensuring fair prices and sustainable livelihoods.

Grassroots-Level Innovations
Grassroots-level innovations involve local solutions developed by farmers themselves, often in response to specific challenges faced in their communities. These innovations are characterized by their adaptability and relevance to local conditions.

- Farmer-led research groups: These groups conduct on-farm trials and experiments to identify effective agricultural practices tailored to local environments. The results are shared among members, promoting knowledge transfer and innovation.
- Local knowledge-sharing networks: These are informal networks of farmers who regularly meet to exchange information, experiences, and solutions. These networks enhance community resilience and the capacity to cope with agricultural challenges.

Challenges and Limitations of Digital Agro-Advisory Services

- Though digital agro-advisory services offer numerous benefits, there exist several challenges and limitations that hinder their widespread adoption and sustainability. These challenges can be widely categorized as technological barriers, socio-economic constraints, and policy and regulatory issues.

Technological Barriers

- **Digital divide:** The uneven distribution of digital infrastructure, particularly in rural and remote areas, poses a significant barrier. Limited access to the Internet, smartphones, and other digital devices prevents many farmers from utilizing digital agro-advisory services. According to Hennessy et al. (2016), this digital divide is a major obstacle to the adoption of technology in agriculture.
- **Digital illiteracy:** Many farmers, especially older ones, may lack the technical skills required to effectively use digital platforms. This lack of digital literacy can result in low adoption rates and ineffective use of available technologies. Magesa et al. (2023) highlight that improving technical literacy among farmers is essential for the success of digital agro-advisory services.
- **Connectivity issues:** Poor network connectivity and frequent power outages in rural areas can disrupt the delivery and reliability of digital services. Chavula (2014) notes that these connectivity issues affect the timeliness and accuracy of the information provided, making it challenging for farmers to rely on digital advisories.
- **Data security and privacy:** Concerns about data security and privacy hinder farming community from using digital services. Building trust with farmers are key challenges for digital service providers. The Food and Agriculture Organization (FAO 2019) emphasizes the importance of addressing data security and privacy issues to encourage the adoption of digital technologies.

Socio-Economic Constraints

- Affordability: Many smallholder farmers cannot afford digital devices and Internet services, which limits their access to digital agro-advisory services. Deichmann et al. (2016) suggest that affordability is a significant barrier that needs to be addressed to increase adoption rates.
- Gender disparities: Women farmers often face greater barriers to accessing digital services due to sociocultural norms. With comparatively lower literacy levels and limited control over financial resources, women are facing challenges for the digital connect. The GSMA (2020) reports that addressing gender disparities is crucial for ensuring inclusive access to digital agro-advisories.
- Trust in technology: Farmers may be skeptical about the reliability and relevance of digital information compared to traditional sources, showcasing minimal to no trust in the information provider. Building trust through consistent and accurate advice is essential for the adoption of digital services. Mittal and Mehar (2016) argue that trust in technology is a key factor influencing farmers' willingness to use digital tools.
- Economic stability: Fluctuations in market prices and economic instability can affect farmers' willingness to invest in digital tools. Ensuring that digital services provide tangible economic benefits can help mitigate this challenge. Fafchamps and Minten (2012) highlight the importance of demonstrating the economic value of digital agro-advisory services to encourage investment.

Policy and Regulatory Issues

- **Regulatory frameworks:** Clear rules for digital agriculture are needed to reduce uncertainty for service providers and users. Good policies should support innovation and protect users' rights. The FAO (2017) highlights the need for comprehensive rules to help adopt digital technologies.
- **Standardization:** Without standardized methods for data handling, there can be inconsistencies and issues with compatibility. Creating standard practices will make digital agro-advisory services more effective. The World Economic Forum (2018) supports developing these standards for better service delivery.
- **Supportive policies:** Governments should create policies that encourage using digital technologies in agriculture. This includes investing in infrastructure, providing subsidies for digital tools, and offering training programs for farmers. Donovan emphasizes that supportive policies are key to the growth of digital agro-advisory services.
- **Public-private collaboration:** Working together, governments, private companies, and nonprofits can scale up digital agro-advisory services. Policies that encourage these partnerships can improve service delivery and reach. The FAO (2019) stresses the importance of such collaborations to promote digital technologies in agriculture.

Future Trends and Opportunities in Digital Agro-Advisory Services

Emerging Technologies (AI, IoT, Big Data)

Both articles highlight the transformative potential of emerging technologies in agriculture:

- **Artificial intelligence (AI)**: AI is revolutionizing agriculture by enabling predictive analytics for weather, pest management, and crop yield optimization (Sonka 2020). Dhanaraju et al. (2022) emphasize AI's role in predictive analytics for disease and pest management, enabling proactive farming practices.
- **Internet of Things (IoT)**: IoT devices, including sensors and actuators, enable real-time monitoring and data collection from fields and equipment (Sonka 2020; Dhanaraju et al. 2022). These devices facilitate precision agriculture and automated systems for irrigation and crop management.
- **Big data**: The collection and analysis of large volumes of data provide valuable insights into soil health, market trends, and consumer preferences (Sonka 2020; Dhanaraju et al. 2022). Big data analytics support decision-making processes for farmers, enhancing productivity and sustainability.

Sonka (2020) discusses the integration of AI, IoT, and big data into agro-advisory services, emphasizing personalized advisory based on data analytics and increased connectivity across the agricultural ecosystem. Dhanaraju et al. (2022) highlight the implementation of IoT-based smart farming practices, demonstrating improvements in crop yields, cost savings, and overall farm management efficiency.

Integration with Sustainable Farming Practices

Digital agro-advisory services (DASs) play a crucial role in facilitating farmers' access to timely and relevant information, which significantly enhances their productivity. These services leverage various digital platforms to disseminate agricultural knowledge, best practices, and real-time advice, thereby bridging the gap between farmers and expert guidance. By utilizing technologies such as mobile applications, SMS, and online portals, DAS provides farmers with critical information on weather forecasts, pest and disease management, crop selection, soil health, and market trends. This access to comprehensive and actionable insights helps farmers make informed decisions, optimize resource use, and ultimately improve their agricultural output.

A prime example of the effectiveness of DAS is the AgriPath project in Grameen Foundation India, which collaborates with Farmbetter India to integrate mobile phone technology in agricultural advisory services.

Farmbetter: Advancing Sustainable Agriculture Through Digital Innovation

Farmbetter is an innovative digital platform that provides sustainable land and crop management practices to farmers worldwide. Other stakeholders are agricultural experts, researchers, and policymakers. It has more elaborately explained content coupled with photographs and videos for better understanding and adoption by their stakeholders. Farmbetter seeks to increase productivity, enhance sustainability, and increase resilience in farming.

WOCAT Global SLM Database

The core source of information for Farmbetter is the WOCAT Global SLM Database (World Overview of Conservation Approaches and Technologies for Sustainable Land Management). This database is maintained by the WOCAT network, a worldwide collaboration of institutions and experts dedicated to promoting sustainable land management. Since its establishment in 1992 by the Swiss Agency for Development and Cooperation (SDC) and the Centre for Development and Environment (CDE) at the University of Bern, Switzerland, WOCAT has expanded to include numerous partners, research institutions, NGOs, and governmental agencies.

Utilizing WOCAT in Farmbetter

Farmbetter enhances its digital agricultural extension services by leveraging extensive knowledge from the WOCAT Global SLM Database. The app provides users with evidence-based, tailored advice on soil conservation, water management, crop rotation, and other sustainable farming techniques drawn from the detailed practices documented in the WOCAT database. This ensures that the recommendations are reliable and relevant, helping farmers implement best practices, improve sustainability, and increase agricultural productivity.

AgriPath: A Research Initiative

AgriPath is committed to expanding sustainable agriculture by identifying and promoting effective and efficient digital advisory services. The initiative aims to empower smallholder farmers—both women and men—with the knowledge and tools necessary to make informed decisions. By encouraging the adoption of sustainable farming methods, AgriPath helps farmers enhance productivity and income while increasing their resilience to climate change. This effort aims to build a more resilient and sustainable agricultural sector, benefiting farmers, their communities, and the environment (https://www.agripath.net/).

Leveraging Potential and Ensuring Sustainability

AgriPath employs a three-pronged strategy to maximize potential and ensure sustainability:

- **Harnessing big data:** Providing farmers and communities with location-specific information on various practices and their benefits to support decision-making and the scaling up of effective solutions.
- **Future-ready app**: The Farmbetter app will be developed to be sustainable and profitable and designed to meet future agricultural needs.
- **Continuous learning**: Encouraging widespread use of the Farmbetter app among farmers and agents, scaling up knowledge-intensive agriculture, and providing insights for other digital advisory services to enhance their platforms.

AgriPath's Implementation in India

In India, AgriPath collaborates with the Grameen Foundation India to conduct randomized control trials in 160 villages in Purvanchal, Uttar Pradesh. These trials, spanning three cropping seasons, include three different models (agent based, hybrid, and self-service) along with a control group. The Farmbetter app, featuring scientifically documented techniques and practices available in Hindi, is integral to these experiments, aiming to improve understanding and adoption of sustainable farming methods. The project started from March 2021 and will conclude in February 2026. Not only in India, but the project is also operative in five nations, two in Asia (India, Nepal) and three in sub-Saharan Africa (Uganda, Tanzania, and Burkina Faso). The team of scientists, researchers, implementing agencies, postdoc, and doc students worldwide are associated with the project.

Participants in the study receive information and support through:

- Agro-advisory services
- Interactive features
- Knowledge repository

This research-based approach helps identify the most effective strategies for improving crop yields, soil health, and farm sustainability. As new evidence emerges, the Farmbetter app will continue to evolve and improve, adapting to meet the needs of farmers worldwide.

Among the digital tools available, mobile phones are particularly effective in conveying messages to farmers. Mobile phones offer unparalleled accessibility and convenience, enabling farmers to receive updates and alerts directly, regardless of their location. Mobile-based DAS can deliver personalized recommendations through SMS and app notifications, ensuring that farmers receive the most relevant information tailored to their specific needs and conditions. Additionally, mobile phones support multimedia content, such as instructional videos and infographics, which can enhance understanding and facilitate the adoption of new practices. The widespread use of mobile phones, even in remote areas, makes them an ideal medium for extending the reach of digital agro-advisory services, empowering farmers with the knowledge they need to boost their productivity and sustain their livelihoods.

Beyond mobile phones, other emerging technologies are also playing critical roles in enhancing SLMPs and overall agricultural practices. Artificial intelligence

(AI) and machine learning algorithms analyze vast amounts of data to provide predictive analytics and early warning systems for pest and disease outbreaks. Internet of Things (IoT) devices, such as soil moisture sensors and weather stations, offer real-time monitoring and precision farming solutions that optimize resource use and improve crop management. Big data analytics facilitate comprehensive farm management by providing actionable insights into crop performance and resource efficiency. Blockchain technology ensures transparency and traceability in supply chains, enhancing trust and compliance with sustainable practices. Remote sensing and geographic information systems (GIS) provide detailed spatial and temporal data for effective land and crop management. Collectively, these technologies contribute to more resilient, efficient, and sustainable agricultural systems, addressing the challenges faced by traditional farming methods and driving the transformation towards a more advanced agricultural future.

Policy Support and Institutional Framework for Enhancing Digital Agro-Advisory Services

A needs-based assessment is essential to ensure that an agro-advisory app meets the specific requirements of farmers. Once this assessment is completed, the app should be redesigned to include several key features aimed at enhancing its utility and accessibility.

Key Features of the Agro-Advisory App

- Multilingual assistance: The app should support multiple languages to cater to farmers from diverse linguistic backgrounds so they can easily understand the content and the context.
- Real-time image capture: Farmers should be able to capture and upload photos within the app for real-time diagnosis for consultation from a subject matter specialist.
- Chatbot availability: A chatbot should be integrated to provide immediate assistance and answer queries.
- Interactive voice response (IVR) system: An IVR system should be included for farmers who are comfortable and have preference to voice interactions.
- Reach to government schemes and subsidies: The app should provide easy access to information about government schemes and subsidies, enabling farmers to avail benefits or connect to the concerned government authority with just one click.
- Repository of sustainable land management practices (SLMPs): A comprehensive repository should provide information on practices such as green manuring, integrated pest management, integrated farming, and organic manuring.
- Market knowledge and insights: Information on current market trends and prices should be available to keep farmers informed.

To ensure the app's success and widespread adoption, it should receive government approval and be supported by a policy framework designed to benefit farmers.

Policy Support for Digital Agro-Advisory Services

- Successful implementation and adoption of digital agro-advisory services depend heavily on strong policy support from the government. The policy framework should focus on the following key areas:
- Investment in digital infrastructure: Governments should improve digital infrastructure, especially in rural and remote areas. This includes expanding Internet connectivity, ensuring stable power supplies, and providing access to affordable digital devices.
- Regulatory guidelines/regulatory policies: Establishing clear and supportive regulatory frameworks is crucial. Policies should address data privacy, cybersecurity, and digital literacy to build trust and encourage adoption among farmers.
- Subsidies and incentives: Providing subsidies for smartphones and data packages can reduce financial barriers for smallholder farmers. Incentives for using digital platforms for agricultural purposes can boost adoption rates.
- Training and capacity building: Governments should invest in training programs to enhance the digital literacy of farmers. This includes workshops, online courses, and field demonstrations to educate farmers on using digital tools effectively.
- Research and development: Encouraging research and development in digital agriculture through grants and collaborations with academic institutions can lead to continuous innovation and improvement of digital agro-advisory services.
- Outreach and awareness programs: Conducting public awareness campaigns to highlight the benefits of digital agro-advisory services can change mindsets and encourage more farmers to adopt these technologies.

Institutional Framework

- A robust institutional framework is essential to support the implementation and sustainability of digital agro-advisory services:
  - **Government agencies:** Ministries of Agriculture and Information Technology should collaborate to oversee the development, implementation, and monitoring of digital agro-advisory services.
  - **Public-private partnerships (PPPs):** Collaborations between government bodies, private-sector companies, and NGOs can enhance the delivery of digital agro-advisory services. PPPs can leverage funding, technology, and grassroots outreach.
  - **Extension services:** Traditional agricultural extension services should be integrated with digital platforms to provide a comprehensive advisory system. Extension agents can facilitate the onboarding of farmers onto digital platforms and offer personalized support.
  - **Farmer cooperatives and associations:** Engaging local farmer cooperatives and associations can promote the app and encourage its use among members.

These organizations can provide feedback to improve the app's features and usability.

- **Research institutions:** Collaboration with agricultural research institutions ensures that the app provides scientifically accurate and up-to-date information. Research institutions can analyze data collected through the app to identify trends and areas for improvement.
- **Monitoring and evaluation:** Establishing a robust monitoring and evaluation framework is essential to assess the impact of digital agro-advisory services. Regular assessments can identify challenges, measure success, and make necessary adjustments.

By combining strong policy support with an effective institutional framework, governments can ensure the successful implementation and widespread adoption of digital agro-advisory services. This will lead to improved agricultural practices, increased productivity, and enhanced livelihoods for farmers, contributing to sustainable agricultural development.

Conclusion

In conclusion, the integration of digital technologies in agriculture, particularly through mobile agro-advisory services, has the potential to revolutionize farming practices and significantly enhance productivity. These services bridge the information gap between farmers and the latest agricultural knowledge, providing timely, accurate, and actionable advice. By offering features such as multilingual support, instant photo capture, chatbot assistance, and interactive voice response, these apps cater to the diverse needs of farmers, making modern agricultural practices more accessible.

The success of such digital platforms relies heavily on a supportive policy and institutional framework. Government approval, investment in digital infrastructure, clear regulatory frameworks, subsidies, incentives, and comprehensive training programs are crucial for widespread adoption. By addressing technological barriers, socio-economic constraints, and policy challenges, these digital tools can empower farmers to adopt sustainable land management practices, improve their productivity, and enhance their livelihoods.

The example of the AgriPath project illustrates how collaboration with technology providers like Farmbetter can create a robust platform that offers a wealth of information on sustainable practices. Such initiatives demonstrate the tangible benefits of digital agro-advisory services in real-world scenarios, highlighting their potential to drive significant positive change in the agricultural sector.

As we move forward, continuous innovation and improvement in digital agriculture, supported by strong policies and public-private partnerships, will be essential. By leveraging emerging technologies like AI, IoT, big data, blockchain, and remote sensing, we can create more efficient, resilient, and sustainable agricultural systems that are well equipped to meet the challenges of the future. Ultimately, the goal is to create an inclusive agricultural ecosystem where every farmer, regardless of their location or resources, can benefit from the advancements in digital technology.

References

AGRIPATH, Grameen Foundation India (n.d.). https://www.agripath.net/

Bayer (2020) Climate FieldView. Retrieved from Bayer Digital Farming

Birner R, Anderson JR (2007) How to make agricultural extension demand-driven? The case of India

Chavula HK (2014) The role of ICTs in agricultural production in Africa. J Dev Agric Econ 6(7):279–289

CropIn (2020) Smart Farming Solutions. Retrieved from CropIn Technology

Davis K, Alex G (2020) Strengthening agricultural extension and advisory systems: Procedures for assessing, transforming, and evaluating extension systems

Deichmann U, Goyal A, Mishra D (2016) Will digital technologies transform agriculture in developing countries? Agric Econ 47(S1):21–33

Dhanaraju M, Chenniappan P, Ramalingam K, Pazhanivelan S, Kaliaperumal R (2022) Smart farming: Internet of Things (IoT)-based sustainable agriculture. Agriculture 12(10):1745. https://doi.org/10.3390/agriculture12101745

Fafchamps M, Minten B (2012) Impact of SMS-Based Agricultural Information on Indian Farmers. World Bank Econ Rev 26(3):383–414

FarmLogs (2021) Digital Farm Management Tools. Retrieved from FarmLogs

Finger R (2023) Digital innovations for sustainable and resilient agricultural systems. Eur Rev Agric Econ 50(4):1277–1309. https://doi.org/10.1093/erae/jbad021

Food and Agriculture Organization (FAO) (2017) Information and Communication Technology (ICT) in agriculture: a report to the G20 agricultural deputies. FAO, Rome

Food and Agriculture Organization (FAO) (2019) E-agriculture in action: blockchain for agriculture: opportunities and challenges. FAO, Bangkok

Fuglie KO et al (2020) The growing role of the private sector in agricultural research and development world-wide. Global Food Secur 10:29–38

Ganeshkumar R, Dilipkumar D (2021) Public-Private Partnerships in Agriculture. 3(10). October 2021

Ghimire R, et al (2021) Agricultural Extension and Advisory Services: What Have We Learned? Where Do We Go from Here?

GSMA (2020) The Mobile Gender Gap Report 2020. GSMA, London

Hennessy T, Läpple D, Moran B (2016) The digital divide in farming: a problem of access or engagement? Appl Econ Perspect Policy 38(3):ppw015. https://doi.org/10.1093/aepp/ppw015

John Deere (2021) Precision Agriculture. Retrieved from John Deere

Kansiime MK, et al (2019) Influence of CABI's Direct2Farm mobile services on farmers' adoption of new practices in India

Magesa M, Jonathan J, Urassa J (2023) Digital literacy of smallholder farmers in Tanzania. Sustainability 15(17):13149. https://doi.org/10.3390/su151713149

Mittal S, Mehar M (2016) Socio-economic factors affecting adoption of modern information and communication technology by farmers in India: analysis using multivariate probit model. J Agric Educ Ext 22(2):199–212

Roy P, Saha P (2022) Digital agriculture – The future of Indian farming. Indian Council Agric Res 3(11):10–14

Sonka ST (2020) Digital technologies, big data, and agricultural innovation. In: Sonka ST (ed) The innovation revolution in agriculture. Springer, Cham, pp 145–164. https://doi.org/10.1007/978-3-030-50991-0_8

World Economic Forum (2018) Innovation with a purpose: the role of technology innovation in accelerating food systems transformation. World Economic Forum, Geneva

Improved Nutrient Management and Fertilizer Optimization in Rice-Based Cropping Systems

7

Jhonsonraju Sankati, Ajay Kumar Mishra, Protik Banerjee, and Pavithra Pidikiti

Introduction

Importance of Nutrient Management in Rice Cultivation

Rice (*Oryza sativa*) is a staple food for more than half of the world's population. Efficient nutrient management is crucial for maximizing rice yields and ensuring food security. Nutrient management encompasses the appropriate use of fertilizers, organic amendments, and crop management practices to optimize the availability and uptake of essential nutrients by rice plants. Standardized fertilizer recommendations do not adequately provide the optimal quantity of nutrients required for agricultural systems, leading to either excessive or insufficient nutrient application. This limitation hampers the efficient utilization of nutrients in crop production (Sanyal et al. 2014).

Challenges in Conventional Fertilizer Use

The use of conventional fertilizers in rice cultivation presents several challenges, which impact both agricultural productivity and environmental sustainability. Conventional fertilizers often lead to an imbalance of nutrients in the soil. Rice crops require a specific balance of nitrogen (N), phosphorus (P), and potassium (K), but traditional fertilizers may not provide these nutrients in the required ratios.

J. Sankati · A. K. Mishra (✉) · P. Pidikiti
International Rice Research Institute- South Asian Regional Centre, Varanasi, Uttar Pradesh, India
e-mail: a.k.mishra@irri.org

P. Banerjee
The University of Texas, San Antonio, TX, USA

A. K. Mishra et al. (eds.), *Key Drivers and Indicators of Soil Health Management*, https://doi.org/10.1007/978-981-97-7564-4_7

Additionally, nutrient-use efficiency is generally low, with significant portions of applied fertilizers not being taken up by the plants, leading to wastage (Cassman et al. 2002). Excessive use of conventional fertilizers can cause environmental pollution. Runoff from fertilized fields can lead to the eutrophication of water bodies, where an overabundance of nutrients, particularly nitrogen and phosphorus, causes excessive growth of algae and other aquatic plants. This can result in the depletion of oxygen in the water, harming aquatic life (Zhang et al. 2014; Craswell 2021). Continuous and intensive use of chemical fertilizers can degrade soil health over time. It can lead to soil acidification, reduction in soil organic matter, and deterioration of soil structure, which ultimately affects crop yields and sustainability (Miao et al. 2011). The application of conventional fertilizers contributes to greenhouse gas emissions of CO_2, CH_4, and particularly nitrous oxide (N_2O) with N inputs, a potent greenhouse gas. This occurs through microbial processes in the soil that transform nitrogen from fertilizers into N_2O (Smith et al. 2007). Hence, there has been a huge increase in global greenhouse gas (GHG) emissions by the agriculture sector, primarily due to the use of synthetic fertilizers and pesticides in recent years, both of which are rapidly expanding. One of the primary causes of environmental pollution, such as eutrophication and GHG emissions, is the injudicious use of fertilizer, particularly nitrogen (Davidson et al. 2014) (Fig. 7.1 and Table 7.1).

Opportunities for Improving Nutrient Management

To address these challenges and maintain long-term environmental and agricultural sustainability, the adoption of precision nutrient management in rice cultivation has become crucial (Parihar et al. 2020; Sharma et al. 2019). The nutrient management (NM), at a specific site, is an approach for dynamic nutrient optimization in rice to overcome the supply and demand of nutrients. The underlying premise of NM is to

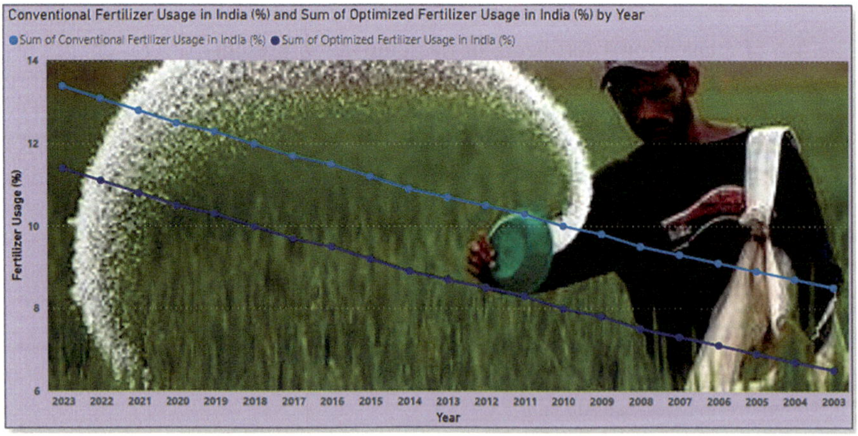

Fig. 7.1 Conventional and optimized fertilizer usage in India (FAO 2023)

Table 7.1 Fertilizer usage and GHG emissions in rice cultivation in Asia and India

Year	Fertilizer usage in Asia (%)	Fertilizer usage in India (%)	GHG emissions from rice cultivation in Asia (Mt CO_2e)	GHG emissions from rice cultivation in India (Mt CO_2e)
2003	10.2	8.5	150	45
2004	10.5	8.7	152	46
2005	10.8	8.9	154	47
2006	11	9.1	156	48
2007	11.2	9.3	158	49
2008	11.5	9.5	160	50
2009	11.8	9.8	162	51
2010	12.1	10	164	52
2011	12.4	10.3	166	53
2012	12.6	10.5	168	54
2013	12.9	10.7	170	55
2014	13.2	10.9	172	56
2015	13.5	11.2	174	57
2016	13.8	11.5	176	58
2017	14.1	11.7	178	59
2018	14.4	12	180	60
2019	14.7	12.3	182	61
2020	15	12.5	184	62
2021	15.3	12.8	186	63
2022	15.6	13.1	188	64
2023	15.9	13.4	190	65

optimize nutrients applied to crops according to appropriate time and rate. The NM allows to calculate the required nitrogen (N), phosphorus (P), and potassium (K) amounts to attain a targeted yield, with additional rules for the timing of fertilizer applications, thereby providing concrete advice to farmers on the rate, sources, and time of fertilizer applications. Nutrient management can improve crop yield and income. This approach helps in enhancing productivity, maintaining soil fertility, and minimizing environmental impacts (Fageria 2007). This approach involves synchronizing fertilizer application with crop needs. Numerous smart tools and decision support systems are available to determine the precise timing and appropriate dosage of primary nutrients tailored to specific sites. These tools rely on continuous monitoring and surveying to optimize nutrient management strategies. To date, there has been limited research on assessing the most effective smart or decision support tool for synchronizing fertilizer application with crop needs while ensuring optimal food production. Therefore, this chapter brings the different nutrient optimization techniques together to better use for transforming the conventional agriculture to regenerative agriculture (Tables 7.2 and 7.3).

Table 7.2 Cultivation area (%) and synthetic fertilizer use percentage in India and Asia

	Rice cultivation		Synthetic fertilizer	
Year	Asia	India	Asia	India
2011	20.2	15.1	17.4	6.1
2012	20.2	15.1	17.4	5.9
2013	20.4	15.5	17.2	4.5
2014	20	15.3	17.4	4.7
2020	19.5	15	17	4.5
2021	19.2	14.8	16.8	4.3
2022	19	14.5	16.7	4.1
2023	18.8	14.3	16.5	4

Source: FAO (2023)

Table 7.3 Comparison of conventional and optimized fertilizer usage and their GHG emissions

Year	Conventional fertilizer usage in India (%)	GHG emissions from conventional rice cultivation in India (Mt CO_2e)	Optimized fertilizer usage in India (%)	GHG emissions from optimized rice cultivation in India (Mt CO_2e)
2003	8.5	45	6.5	42
2004	8.7	46	6.7	43
2005	8.9	47	6.9	44
2006	9.1	48	7.1	45
2007	9.3	49	7.3	46
2008	9.5	50	7.5	47
2009	9.8	51	7.8	48
2010	10	52	8	49
2011	10.3	53	8.3	50
2012	10.5	54	8.5	51
2013	10.7	55	8.7	52
2014	10.9	56	8.9	53
2015	11.2	57	9.2	54
2016	11.5	58	9.5	55
2017	11.7	59	9.7	56
2018	12	60	10	57
2019	12.3	61	10.3	58
2020	12.5	62	10.5	59
2021	12.8	63	10.8	60
2022	13.1	64	11.1	61
2023	13.4	65	11.4	62

Source: FAO (2023)

Understanding Nutrient Requirements in Rice

Macronutrients and Micronutrients Essential for Rice Growth

Rice growth relies on a balanced supply of essential macronutrients and micronutrients. Nitrogen (N) is vital for amino acids, proteins, nucleic acids, and chlorophyll, promoting vegetative growth and enzyme synthesis; its deficiency leads to pale green leaves and stunted growth (Wang et al. 2022). Phosphorus (P) is crucial for energy transfer, root development, and flowering, with deficiencies causing dark green leaves and poor root growth (Zhu et al. 2023). Potassium (K) regulates osmotic balance, enzyme activation, and photosynthesis, with deficiencies resulting in yellowing leaf margins and weak stems (Liu et al. 2023a, b). Sulfur (S) is a component of amino acids and chlorophyll, necessary for protein synthesis, with deficiencies causing yellowing of young leaves (Zhao et al. 2023). Magnesium (Mg) is central to chlorophyll and photosynthesis, with deficiencies leading to interveinal chlorosis (Chen et al. 2022). Calcium (Ca) is essential for cell wall structure and membrane integrity, with deficiencies causing necrosis of young leaf tips (Xu et al. 2022). Micronutrients like iron (Fe), manganese (Mn), zinc (Zn), copper (Cu), boron (B), molybdenum (Mo), and chlorine (Cl) are also critical. Iron is crucial for chlorophyll synthesis. Liu et al. (2023a, b) emphasized the importance of iron for improving chlorophyll content and photosynthetic capacity in rice. Manganese is essential for photosynthesis and nitrogen metabolism. Wang et al. (2023a, b) highlighted that manganese application enhances disease resistance and enzyme activities in rice plants. Zinc is required for enzyme function. Recent research by Sharma et al. (2023) indicates that zinc fertilization improves rice yield and grain nutritional quality. Copper is needed for photosynthesis and respiration. Yang et al. (2023) demonstrated that copper application enhances antioxidant enzyme activities and mitigates oxidative stress in rice. Boron is essential for cell wall formation. According to Huang et al., boron application improves pollen viability and grain set in rice. Molybdenum is crucial for nitrogen fixation. Research by Zhang et al. suggests that molybdenum supplementation enhances nitrogen-use efficiency in rice. Chlorine is essential for osmosis and ionic balance. Recent studies indicate that chlorine plays a role in disease resistance and overall plant vigor in rice. Deficiencies in these micronutrients lead to various symptoms like interveinal chlorosis, necrotic spots, and poor growth. Continuous research and advanced nutrient management are essential to optimize rice productivity and meet global food demands.

Soil Fertility Assessment Techniques (Fertilizer Optimization Techniques)

Soil Testing and Nutrient Mapping

Assessing soil fertility is vital for sustaining agricultural productivity and environmental health. Common techniques include soil sampling, pH measurement, nutrient analysis, assessment of organic matter content, cation-exchange capacity (CEC),

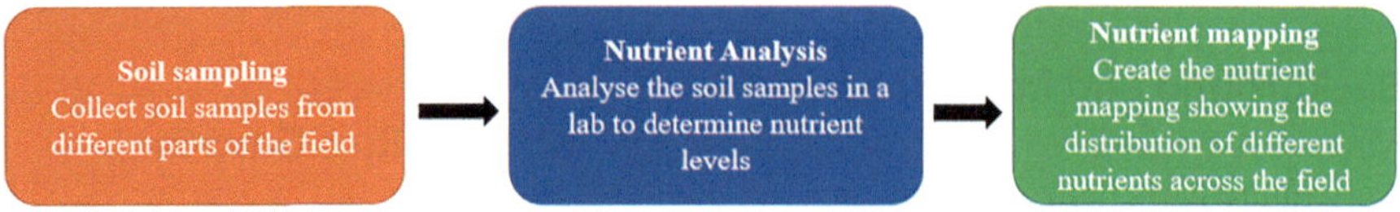

Fig. 7.2 Process of nutrient mapping

microbial analysis, soil respiration, texture analysis, electrical conductivity (EC), and plant tissue analysis (Brady and Weil 2016; Gee and Bauder 1986; Nelson and Sommers 1996; Vance et al. 1987). These techniques offer insights into soil physical, chemical, and biological properties, which collectively influence plant growth and soil health. Advanced methods, such as X-ray fluorescence (XRF) spectroscopy, near-infrared reflectance spectroscopy (NIRS), and DNA-based techniques, provide deeper insights into soil characteristics (Viscarra Rossel et al. 2016). Additionally, electron microscopy, isotope analysis, soil electrical resistivity imaging (ERI), laser-induced breakdown spectroscopy (LIBS), soil microarray analysis, and digital soil mapping (DSM) offer advanced capabilities for detailed soil analysis, facilitating precise soil management strategies. These methods enable comprehensive assessments of soil fertility parameters, contributing to informed decision-making in agriculture and environmental management (Fig. 7.2).

Remote Sensing Technologies for Nutrient Assessment

Remote sensing technologies offer a comprehensive suite of tools for assessing nutrient status in agricultural fields, providing nondestructive and spatially explicit information crucial for precision nutrient management. By leveraging various sensors mounted on satellites, aircraft, drones, or ground-based platforms, remote sensing captures multispectral or hyperspectral imagery, which is then processed and analyzed to derive indicators of nutrient status. These technologies include methods such as the normalized difference vegetation index (NDVI) and ratio vegetation index (RVI), which correlate with vegetation greenness and density, serving as proxies for biomass and nitrogen content (Clevers et al. 2017). Additionally, thermal infrared sensors monitor canopy temperature, hyperspectral imaging estimates leaf nitrogen concentration, and machine learning algorithms integrate multispectral data for nutrient deficiency classification (Gitelson et al. 2003). Such technologies enable the identification of nutrient deficiency symptoms and spatial patterns across agricultural landscapes, facilitating informed decisions regarding nutrient management practices and interventions.

Precision Agriculture Approaches

Chlorophyll Meters

SPAD Meter

Handheld chlorophyll meters offer a rapid and precise method for assessing nitrogen (N) needs in plants. These meters, exemplified by the Soil Plant Analysis Development (SPAD502®), gauge chlorophyll content by measuring light transmittance or reflectance. The SPAD502® employs light-emitting diodes (LEDs) emitting red and infrared light, instantly estimating leaf N status based on chlorophyll content. It measures the relative difference in crop N status, detecting N stress early on, even before visible symptoms appear (Francis and Piekielek 1999). Research has focused on two areas: establishing linear relationships between SPAD readings and leaf N content and determining thresholds for top-dressing N application based on SPAD values (Balasubramanian et al. 1999). Despite variations in crop types, these meters provide a scientific and accessible means to optimize N management, contributing to enhanced crop productivity and resource efficiency (Fig. 7.3).

Leaf Color Chart (LCC)

Leaf color chart (LCC) offers a simple and cost-effective method for assessing plant nitrogen (N) status visually. Embedded with green plastic chips ranging from yellowish to dark green, LCC provides a continuum from N deficiency to excess (Fischer et al. 2004; Pasuquin et al. 2004). Unlike chlorophyll meters, LCC measures leaf greenness and associated leaf N through visual comparison of light reflection (Yang et al. 2003). LCCs serve as practical alternatives, especially for resource-limited farmers in countries like India (Bijay-Singh et al. 2002). Developed by various institutions, including the International Rice Research Institute (IRRI) and Zhejiang Agricultural University, LCCs aid in site-specific N management for crops like rice, wheat, and maize (Witt et al. 2005). They facilitate two approaches: fixed splitting pattern and real-time adjustment based on critical LCC values. Optical sensors, another technology, measure visible and near-infrared spectral responses to detect N stress, complementing LCC and chlorophyll meters (Peñuelas et al. 1994; Ma et al. 1996; Raun et al. 2001) (Fig. 7.4).

Soil Modeling and Nutrient Recommendation

Integrating soil modeling with fertilizer recommendation systems revolutionizes precision agriculture, offering tailored guidance for optimal fertilizer application. Drawing on comprehensive input data encompassing soil characteristics, climate, crop specifics, and management practices, these systems leverage sophisticated models to simulate soil processes and crop responses (Basso et al. 2009). Through calibration and validation against field measurements, these models ensure accuracy and reliability in predicting soil behavior and nutrient dynamics. Subsequently,

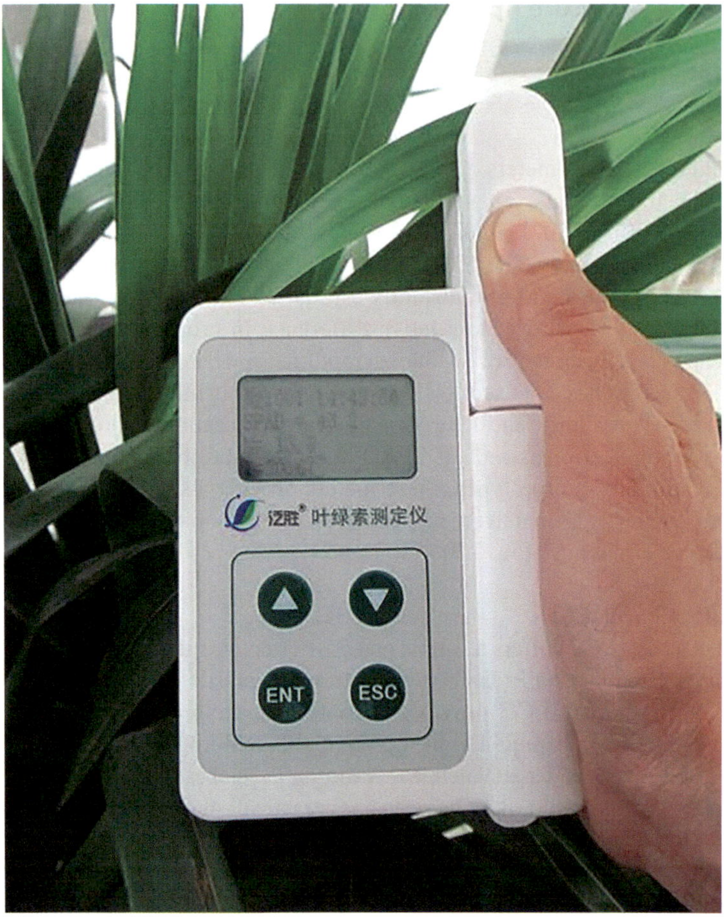

Fig. 7.3 SPAD meter

fertilizer prescription maps are generated, delineating spatial variability in fertilizer requirements within fields, thus enabling variable-rate fertilizer application (Li et al. 2018). These maps, often integrated into decision support systems (DSSs), provide farmers with user-friendly interfaces for accessing and interpreting recommendations. The DSS, enriched with economic analyses, empowers farmers to optimize fertilizer investments and maximize returns. Moreover, continuous monitoring and adaptive management allow for real-time adjustments to recommendations based on evolving conditions, ensuring sustained effectiveness (Pan et al. 2020). Regular evaluation and refinement of these systems further enhance their accuracy and relevance for on-the-ground decision-making in agriculture. Through this integrated approach, farmers can make informed decisions, promoting efficient nutrient management, enhanced crop productivity, and sustainability.

Fig. 7.4 Leaf color chart

Table 7.4 State nutrient recommendation in India for rice crop

State	Rice ecology	Doses of nutrient applied (N-P-K, kg/ha)	References
Andhra Pradesh	**Irrigated lowland**	100–60–40	ICAR-NRRI
Bihar	Rainfed upland	60–30–30	Singh et al. (2020)
Chhattisgarh	Lowland	100–60–40	Shrivastava (2019)
Karnataka	Irrigated lowland	100–50–50	Mowo et al. (1993)
Kerala	Rainfed lowland	90–45–45	ICAR (2020)
Odisha	Irrigated lowland	80–40–40	IRRI (2013)
Tamil Nadu	Rainfed lowland	100–50–50	ICAR (2020)
Uttar Pradesh	Irrigated lowland	120–60–60	ICAR (2020)

Crop-Specific Nutrient Requirements (Tables 7.4 and 7.5)

Table 7.5 State nutrient recommendations for different rice ecosystems and different practices in India

State	Rice ecosystem	RDF (N-P-K, kg/ha)	SSNM (N-P-K, kg/ha)	INM (N-P-K, kg/ha)
Punjab	Irrigated	120–60–40 + rice straw compost @ 6 t/ha	130–60–50	90–60–40 + FYM @ 15 t/ha
Haryana	Irrigated	150–60–40 + rice straw compost @ 6 t/ha	160–60–50	120–60–40 + FYM @ 15 t/ha
Tamil Nadu	Irrigated	150–50–50 + green manure @ 5 t/ha	140–60–60	100–50–50 + vermicompost @ 5 t/ha
West Bengal	Rainfed	100–50–50 + FYM @ 5 t/ha	110–50–50	80–40–40 + poultry manure @ 3 t/ha
Odisha	Rainfed	80–40–40 + FYM @ 5 t/ha	90–40–40	60–30–30 + green manure @ 5 t/ha
Andhra Pradesh	Irrigated	120–60–60 + FYM @ 5 t/ha	130–60–60	100–50–50 + FYM @ 10 t/ha
Bihar	Rainfed	100–60–40 + FYM @ 3 t/ha	110–60–40	80–40–40 + FYM @ 5 t/ha
Karnataka	Irrigated upland	90–45–45 + vermicompost @ 5 t/ha	110–55–55	80–40–40 + FYM @ 10 t/ha
Uttar Pradesh	Irrigated	150–60–60 + FYM @ 3 t/ha	160–60–70	120–60–50 + FYM @ 5 t/ha

Site-Specific Nutrient Management (SSNM) in Rice

Principles of SSNM

Site-specific nutrient management (SSNM) represents a transformative approach in rice cultivation, moving away from traditional blanket fertilizer applications towards a more precise, field-specific nutrient management strategy. This method aims to optimize nutrient supply based on the specific needs of rice crops at various growth stages, thereby enhancing crop productivity and nutrient-use efficiency while minimizing environmental impacts. Traditional fertilizer practices often result in inefficient nutrient use, leading to suboptimal yields, nutrient wastage, and environmental issues such as water pollution and greenhouse gas emissions. SSNM addresses these challenges by considering site-specific factors like soil properties, crop varieties, climatic conditions, and management practices to ensure that nutrients are applied in the right amount, at the right time, and in the right place (Jat et al. 2013).

Components of SSNM

SSNM incorporates several key components that leverage advanced technologies to facilitate precise nutrient management. Geographic information system (GIS), remote sensing, soil testing, and data analytics are crucial in mapping soil variability, monitoring crop health in real time, and dynamically adjusting nutrient applications. These precision agriculture tools enable farmers to make informed decisions that improve yield potential and reduce environmental impact (Schmidt et al. 2021). Soil testing to assess nutrient levels and pH is fundamental, allowing for the creation of detailed nutrient management plans tailored to specific field conditions. Understanding crop nutrient requirements at different growth stages ensures that nutrients are supplied precisely when needed, enhancing growth and yield while minimizing waste. SSNM also emphasizes sustainable practices, such as using organic amendments, crop rotation, and conservation tillage, to improve soil health and long-term agricultural productivity, reducing dependency on chemical fertilizers and enhancing nutrient-use efficiency (Buresh et al. 2018).

Case Studies Demonstrating SSNM Efficacy

Numerous studies have demonstrated the efficacy of SSNM in various rice-growing regions. A study in Tamil Nadu, India, showed that SSNM significantly increased rice yields by 0.7–1.2 tons/ha compared to conventional practices, enhanced nutrient-use efficiency by 18–22%, and reduced environmental impacts by minimizing nutrient losses (Jat et al. 2013). This case study highlights the effectiveness of SSNM in enhancing rice productivity, improving nutrient-use efficiency, and mitigating environmental impacts. Another study in Punjab, India, evaluated SSNM in a high-intensity farming area with diverse soil types and climatic conditions. The implementation of SSNM led to a 10–15% increase in rice yields and improved nitrogen-use efficiency by 20–25%, with a significant reduction in nitrogen leaching and greenhouse gas emissions (Singh et al. 2021). This study underscores the adaptability and effectiveness of SSNM across diverse agroecological contexts, validating its potential for widespread adoption.

The principles of SSNM revolve around considering site-specific factors, optimizing the timing and application of nutrients, and integrating precision agriculture and sustainable practices. Continuous monitoring and adjustment of nutrient applications during the growing season, along with incorporating organic sources of nutrients, further enhance the effectiveness of SSNM. By adopting these practices, farmers can achieve higher crop productivity, improved nutrient-use efficiency, and reduced environmental impact, contributing to sustainable and economically viable agricultural systems (Wang et al. 2023a, b; Lal et al. 2021a, b) (Fig. 7.5).

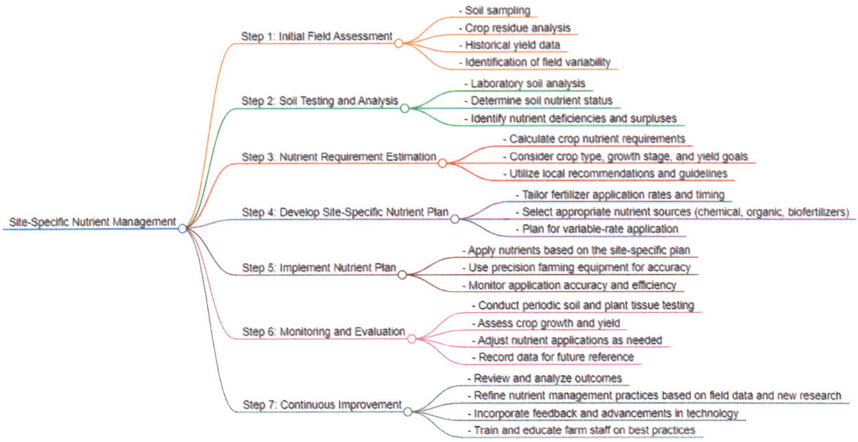

Fig. 7.5 Flowchart illustrating the steps involved in site-specific nutrient management

Integrated Nutrient Management (INM)

Concept and Principles of Integrated Nutrient Management (INM)

Integrated nutrient management is a modified nutrient management technique with multifarious benefits, wherein a combination of all possible sources of plant nutrients is used in a crop nutrition package. INM is a sustainable approach that integrates chemical fertilizers with organic matter, biofertilizers, and crop residues to optimize soil fertility and crop nutrition (Roy et al. 2020). Unlike traditional methods, INM recognizes the complexity of soil systems and aims to balance nutrient supply to enhance both soil health and crop productivity (Chen et al. 2021). The principles of INM focus on maintaining a balance of nutrients, promoting soil health, tailoring practices to specific local conditions, and ensuring environmental sustainability (Luo et al. 2021; Hossain et al. 2022; Pathak et al. 2020). The objectives of INM include boosting crop productivity, maintaining soil fertility for future generations, supporting environmental health, and ensuring economic viability for farmers. By leveraging various nutrient sources and promoting their synergistic use, INM fosters healthier soils, more productive crops, and a resilient agricultural ecosystem, essential for meeting the food security needs of a growing population while preserving the environment for future generations.

Components of INM

INM comprises several key components that collectively improve soil fertility, enhance crop productivity, and promote sustainable agricultural practices. These components include the strategic use of chemical fertilizers, organic manures, biofertilizers, crop residues, soil amendments, crop rotation, cover cropping, precision farming

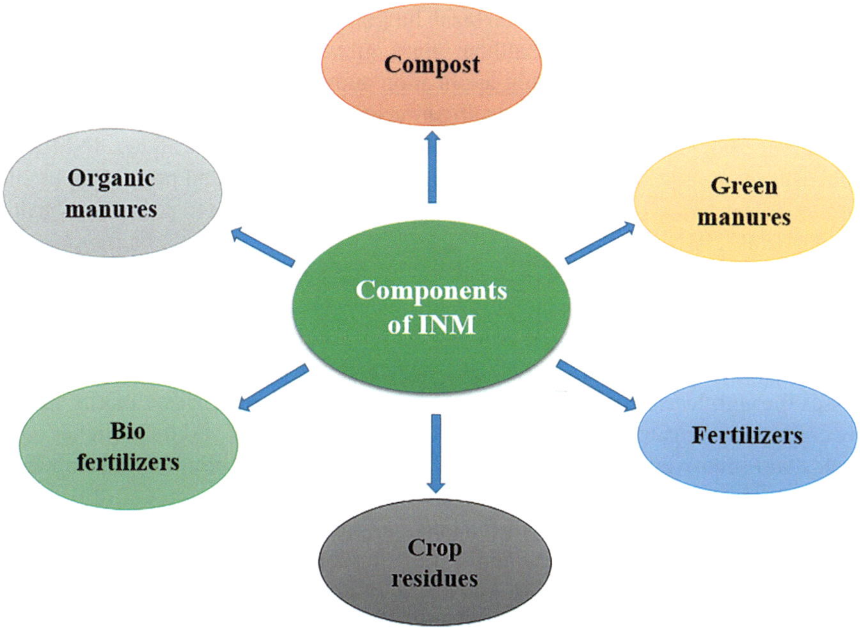

Fig. 7.6 Components of INM

techniques, and farmer education and extension services. Chemical fertilizers provide immediate and concentrated nutrients like nitrogen, phosphorus, and potassium, applied precisely to meet specific crop and soil needs, thus preventing imbalances and minimizing environmental pollution. Organic manures, such as compost and vermicompost, improve soil structure, enhance moisture retention, and stimulate microbial activity, supporting long-term soil fertility and reducing dependence on synthetic fertilizers (Lal 2021). Biofertilizers, including nitrogen-fixing bacteria and mycorrhizal fungi, enhance nutrient availability and uptake, improve soil health, and bolster crop resilience to diseases and stress conditions (Singh et al. 2019a, b). Crop residues add organic matter, improve physical properties, and provide nutrients, supporting soil organic carbon levels, boosting microbial activity, and facilitating nutrient cycling within the agroecosystem (Sharma et al. 2020) (Fig. 7.6).

Benefits of INM Adoption and Sustainable Rice Production Systems

Integrated nutrient management (INM) is crucial for sustainable rice production, integrating organic and biological inputs with precision nutrient management strategies to boost productivity and ensure long-term sustainability. Organic inputs, such as composts and green manures, enhance soil health by improving structure, water retention, and microbial activity (Ghosh et al. 2019). The use of biofertilizers, like nitrogen-fixing bacteria and mycorrhizal fungi, reduces dependency on synthetic fertilizers while promoting healthier plant growth (Singh and Reddy 2020). Precision techniques, such as split application and site-specific management, optimize nutrient utilization efficiency, mitigating environmental impacts such as nutrient runoff and water contamination (Dobermann and Cassman 2020). INM also enhances economic sustainability by minimizing input costs and promoting biodiversity, fostering environmental sustainability, and contributing to ecosystem health (Yadav et al. 2021).

The benefits of INM extend beyond soil health and productivity. By integrating organic inputs and biofertilizers, INM rejuvenates soil fertility, replenishes essential nutrients, and fosters microbial activity, thereby improving soil structure, water retention, and aeration (Lal 2021). It also enhances nutrient-use efficiency, reducing losses through leaching, volatilization, and runoff, which improves nutrient uptake by crops and mitigates environmental pollution. This method results in higher crop yields and improved quality, as the orchestrated nutrient supply facilitates optimal plant growth and enriches the nutritional profile of crops (Singh et al. 2019a, b). Additionally, INM mitigates environmental pollution by curbing the excessive use of chemical fertilizers, thereby preventing nutrient runoff and water pollution (Pathak et al. 2020).

Several studies conducted worldwide have demonstrated the efficacy of INM in significantly improving soil health and crop yields while reducing greenhouse gas emissions and other related problems. The increase in crop yields due to the adoption of INM over conventional nutrient management ranges from 1.3% to 66.5% across major cropping systems. The integration of organic manure and residue retention in INM significantly improves soil aggregates and microbiota (Yadvinder-Singh et al. 2004a, b). Most studies indicate a notable increase in overall soil quality, with lower bulk density, higher porosity, and increased water-holding capacity.

This comprehensive approach aligns with regenerative agriculture principles, emphasizing restoring soil health, enhancing biodiversity, and building resilient farming systems. By incorporating practices such as crop rotation, cover cropping, and reduced tillage, INM supports the regenerative agriculture model, fostering a more sustainable and productive agricultural ecosystem. Practicing INM enhances soil health and crop productivity while decreasing environmental pollution, greenhouse gas emissions, and production costs. This integration within regenerative agriculture frameworks offers a holistic solution to modern agricultural challenges, ensuring long-term sustainability and resilience in farming systems (Jat et al. 2022).

Enhancing Nutrient-Use Efficiency

Timing and Method of Fertilizer Application

Optimizing the timing and method of fertilizer application is crucial for enhancing nutrient-use efficiency (NUE) in rice crops, ensuring that nutrients are available when plants need them most and reducing losses through leaching, volatilization, or runoff. Split application involves dividing the fertilizer into multiple doses throughout the growing season to match the nutrient supply with the crop's demand, significantly increasing NUE and yield. Typically, nitrogen is applied in three splits: basal application at planting, at the tillering stage, and at the panicle initiation stage (Singh et al. 2017). Deep placement of fertilizers, particularly nitrogen, involves placing the fertilizer 7–10 cm below the soil surface, reducing nitrogen losses due to volatilization and runoff and enhancing NUE. Research indicates that deep placement of urea improved nitrogen recovery by 15–20% compared to surface broadcasting (Thapa et al. 2016). Additionally, synchronizing fertilizer application with irrigation events can improve NUE, as applying nitrogen fertilizer before scheduled irrigation ensures that nutrients are readily available to rice plants and less likely to be lost through volatilization (Fageria et al. 2011).

Use of Controlled-Release Fertilizers

Controlled-release fertilizers (CRFs) are designed to release nutrients gradually over time, matching nutrient release with the crop's uptake pattern, thus improving NUE by reducing nutrient losses and providing a consistent supply of nutrients. CRFs can be coated with materials such as sulfur, polymers, or resins to control the rate of nutrient release. Research demonstrates that polymer-coated urea can enhance NUE by 30–50% compared to conventional urea (Shoji et al. 2001). In rice cultivation, CRFs have been shown to reduce the frequency of fertilizer applications and improve yield, with sulfur-coated urea increasing rice yield by 10–15% and reducing nitrogen leaching losses (Chien et al. 2009). By reducing the need for multiple applications and minimizing nutrient losses, CRFs also contribute to a lower environmental footprint, which is particularly important in rice paddies where nitrogen losses can lead to significant environmental issues such as water pollution and greenhouse gas emissions (Akiyama et al. 2010).

Foliar Application and Fertigation Techniques

Foliar application and fertigation are advanced nutrient delivery techniques that can significantly enhance NUE in rice crops. Foliar application involves spraying nutrient solutions directly onto the leaves, allowing for the rapid correction of nutrient deficiencies, and providing nutrients during critical growth stages. This method has been shown to improve rice yield and NUE, especially in soils with poor nutrient availability (Kannan 2010). Fertigation integrates fertilization with irrigation, delivering nutrients directly to the root zone, ensuring a uniform distribution of nutrients and water, and enhancing nutrient uptake efficiency. Studies have shown that fertigation in rice crops can increase nitrogen-use efficiency by 20–30% compared to conventional methods (Phongpan et al. 2000). The integration of fertigation with precision agriculture techniques, such as soil moisture sensors and automated irrigation systems, further enhances NUE by allowing for the precise application of water and nutrients based on real-time soil and crop needs, reducing wastage, and improving productivity (Balasubramanian et al. 2004).

Future Directions and Research Needs

Emerging Technologies for Nutrient Management

Emerging technologies hold immense potential for revolutionizing nutrient management in agricultural systems, including rice-based systems. Some of the key emerging technologies include precision agriculture, remote sensing, and digital farming tools. Precision agriculture utilizes advanced technologies such as GPS, sensors, and drones to precisely manage inputs like fertilizers, water, and pesticides, thereby optimizing resource use and minimizing environmental impacts (Sui et al. 2021). Remote sensing techniques, including satellite imagery and aerial drones, provide valuable data on crop health, soil moisture, and nutrient status, enabling farmers to make informed decisions about nutrient application and management (Doraiswamy et al. 2020). Digital farming tools, such as mobile apps and decision support systems, offer real-time recommendations for nutrient management based on weather forecasts, soil testing results, and crop growth stages, facilitating more efficient and sustainable farming practices (He et al. 2021).

Policy Interventions to Promote Sustainable Practices

Policy interventions play a crucial role in promoting sustainable nutrient management practices in rice-based agricultural systems. Governments can implement policies that incentivize the adoption of practices such as integrated nutrient management, organic farming, and conservation agriculture. Financial incentives, subsidies, and tax breaks can encourage farmers to adopt sustainable practices that enhance soil fertility, reduce chemical inputs, and mitigate environmental impacts (Kaur et al. 2020). Moreover, regulatory measures such as restrictions on the use of certain fertilizers and pesticides, establishment of nutrient management guidelines, and enforcement of environmental regulations can help ensure compliance with sustainable practices and safeguard natural resources for future generations (Singh et al. 2019a, b).

Research Priorities for Advancing Nutrient Management in Rice-Based Systems

Several research priorities are essential for advancing nutrient management in rice-based agricultural systems. These include developing innovative nutrient management strategies tailored to diverse agroecological conditions and farming systems. Research is needed to optimize nutrient application rates, timing, and placement to maximize crop yields while minimizing environmental impacts (Kumar et al. 2021).

Furthermore, there is a need for interdisciplinary research that integrates agronomic, ecological, and socio-economic perspectives to develop holistic solutions for sustainable nutrient management. Long-term field trials, modeling studies, and participatory research approaches can provide valuable insights into the effectiveness and feasibility of different nutrient management strategies in diverse contexts (Mishra et al. 2020a, b; Mishra et al. 2021).

Role of Conservation Agriculture in Nutrient Management

Conservation agriculture (CA) plays a crucial role in nutrient management in rice cropping systems by enhancing soil health and improving nutrient cycling. CA practices, such as minimal soil disturbance, permanent soil cover, and crop rotation, contribute to maintaining and enhancing soil organic matter and microbial activity. These practices reduce soil erosion, increase water infiltration, and improve soil structure, which are essential for efficient nutrient uptake by rice plants (Kumar et al. 2022a). The improved soil health under CA leads to better root growth and nutrient access, reducing the need for synthetic fertilizers and enhancing nutrient-use efficiency (NUE). Studies have shown that CA practices can reduce nutrient losses through leaching and runoff, thereby maintaining soil fertility and contributing to sustainable rice production. For instance, research has demonstrated that implementing CA in rice-based systems can improve NUE by 15–20% compared to conventional tillage systems (Jat et al. 2020; Mishra et al. 2021; Kumar et al. 2022b).

Regenerative agriculture (RA), which shares several principles with CA, further enhances nutrient management in rice cropping systems by focusing on restoring and regenerating soil health. RA practices include diverse crop rotations, cover cropping, and integrated livestock management, which collectively enhance soil nutrient cycling and organic matter content. These practices increase the biological activity in the soil, leading to improved nutrient availability and uptake by rice plants. The synergy between CA and RA creates a resilient agroecosystem where nutrient inputs are optimized, and natural nutrient cycles are restored, reducing dependence on external fertilizers. Recent studies highlight that integrating CA and RA practices in rice farming can significantly enhance soil carbon sequestration, boost microbial activity, and improve overall soil fertility, leading to sustainable and productive rice systems (Lal et al. 2021a, b). This integrated approach not only supports high yields but also aligns with sustainable agricultural goals by promoting environmental health and reducing the carbon footprint of rice production.

Conclusion

Summary of Key Findings

The research on nutrient optimization in rice has highlighted several significant findings. Key among them is the critical role of precision agriculture and site-specific nutrient management (SSNM) in enhancing crop productivity and nutrient-use efficiency. SSNM tailors nutrient application based on field-specific conditions such as soil properties, crop variety, and local climate, resulting in improved yields and reduced environmental impact compared to traditional blanket fertilizer applications. Case studies have shown substantial yield increases (0.7–1.2 tons/ha in Tamil Nadu and 10–15% in Punjab) and improved nutrient-use efficiency (18–22% and 20–25%, respectively) under SSNM practices. Furthermore, conservation agriculture (CA) and regenerative agriculture (RA) practices significantly contribute to nutrient management by enhancing soil health and nutrient cycling, reducing the need for synthetic fertilizers, and improving overall sustainability.

Implications for Sustainable Rice Production

The findings have profound implications for sustainable rice production. By adopting precision agriculture and SSNM, farmers can achieve higher crop productivity with optimized nutrient use, minimizing environmental impacts such as nutrient runoff and greenhouse gas emissions. Conservation and regenerative agricultural practices enhance soil health, leading to long-term sustainability of rice farming systems. These practices help in maintaining soil organic matter, reducing soil erosion, and improving water infiltration, which are essential for sustainable rice production. Additionally, integrating these practices with digital farming tools can facilitate real-time decision-making, further improving the efficiency and sustainability of nutrient management.

Call to Action for Stakeholders

For stakeholders, including policymakers, researchers, and farmers, the call to action is clear. Policymakers should create and enforce regulations that promote sustainable nutrient management practices, such as providing incentives for adopting SSNM, CA, and RA practices. Researchers need to focus on developing innovative, site-specific nutrient management strategies and further integrating advanced technologies like GIS, remote sensing, and digital tools into farming practices. Farmers should embrace these advanced and sustainable practices to improve productivity and environmental sustainability. Collaborative efforts among all stakeholders are essential to transition towards a more sustainable rice production system, ensuring food security and environmental conservation.

References

Akiyama H, Yan X, Yagi K (2010) Evaluation of effectiveness of enhanced-efficiency fertilizers as mitigation options for N2O and NO emissions from agricultural soils: meta-analysis. Glob Chang Biol 16(6):1837–1846

Balasubramanian V, Sie M, Hijmans RJ, Otsuka K (1999) Increasing the productivity of intensive rice systems through site-specific nutrient management. Field Crop Res 64(3):209–229
Balasubramanian V, Morales AC, Cruz RT, Abdulrachman S (2004) On-farm adaptation of knowledge-intensive nitrogen management technologies for rice systems. Nutr Cycl Agroecosyst 67(1):1–18
Basso B, Antle J, Stöckle C (2009) Land use and soil management effects on regional carbon sequestration and greenhouse gas emissions. Ann N Y Acad Sci 1167(1):1–30
Bijay-Singh, Khurana HS, Malik RK (2002) Evaluation of leaf colour chart for managing nitrogen fertilizer in rice. Field Crop Res 74(1):93–101
Brady NC, Weil RR (2016) The nature and properties of soils. Pearson, London
Buresh RJ, Pampolino MF, Witt C, Dobermann A (2018) Nutrient Expert: a tool to optimize nutrient management for improved productivity, profitability, and environmental sustainability. Comput Electron Agric 153:69–77
Cassman KG, Dobermann A, Walters DT (2002) Agroecosystems, nitrogen-use efficiency, and nitrogen management. Ambio 31(2):132–140. https://doi.org/10.1579/0044-7447-31.2.132
Chen J, He F, Zhang X, Zhang S (2021) Advances in biofertilizers and their application in agriculture. J Soil Sci Plant Nutr 21(4):2450–2462
Chen X, Gao L, Zhang Y (2022) Magnesium supplementation enhances chlorophyll content and photosynthetic efficiency in rice. J Plant Physiol 240:153–164
Chien SH, Prochnow LI, Cantarella H (2009) Recent developments of fertilizer production and use to improve nutrient efficiency and minimize environmental impacts. Adv Agron 102:267–322
Clevers JG, Kooistra L, van den Brande MM (2017) Using Sentinel-2 data for retrieving LAI and leaf chlorophyll content estimates for vegetation: a comparison of methods in two landscapes. Sensors 17(7):1581
Craswell E (2021) Fertilizers and nitrate pollution of surface and ground water: an increasingly pervasive global problem. SN Appl Sci 3(4):518. https://doi.org/10.1007/s42452021-04521-8
Davidson EA, Galloway JN, Millar N, Leach AM (2014) N-related greenhouse gases in North America: innovations for a sustainable future. Curr Opin Environ Sustain 9-10:1–8. https://doi.org/10.1016/j.cosust.2014.07.003
Dobermann A, Cassman KG (2020) Plant nutrient management for enhanced productivity in intensive grain production systems. Plant Soil 336(1–2):289–308
Doraiswamy PC, Akhmedova SA, Wang Z, Cosh MH (2020) Applications of remote sensing in agriculture: a review. Remote Sens 12(18):3025
Fageria NK (2007) Green manuring in rice cropping systems. J Plant Nutr 30(5–6):691–720
Fageria NK, Baligar VC, Jones CA (2011) Growth and mineral nutrition of field crops. CRC Press, Hoboken, NJ
FAO (2023) The Food and Agriculture Organization provides updates on agricultural practices and fertilizer usage
Fischer KS, Fukai S, Kumar A (2004) Leaf colour charts for identifying rice nitrogen status. Field Crop Res 89(2–3):299–307
Francis DD, Piekielek WP (1999) Use of a chlorophyll meter to monitor nitrogen status and schedule fertigation for corn. J Plant Nutr 22(3):563–573
Gee GW, Bauder JW (1986) "Particle-size analysis". Methods of Soil Analysis. Part 1. Physical and Mineralogical Methods. Soil Science Society of America
Ghosh S, Wilson B, Ghoshal S, Senapati N, Mandal B (2019) Organic amendments influence soil quality and carbon sequestration in the Indo-Gangetic Plains of India. Agric Ecosyst Environ 266:148–159
Gitelson AA, Viña A, Arkebauer TJ, Rundquist DC, Keydan G, Leavitt B (2003) Remote estimation of leaf area index and green leaf biomass in maize canopies. Geophys Res Lett 30(5):1248
He L, Cui Z, Zhu Y, Zhang F, Chen X (2021) Digital farming: current status and future perspectives. Comput Electron Agric 180:105964
Hossain M, Haque M, Rahman M (2022) Context-specific integrated nutrient management practices for sustainable agriculture. Agric Syst 193:103256

ICAR (2020) National Recommendations for Rice Fertilization. Retrieved from https://sciencedirect.com
International Rice Research Institute (2013) Rice Cultivation Practices and Recommendations. Retrieved from https://books.irri.org
Jat ML, Satyanarayana T, Majumdar K, Pampolino M, Johnston AM, Gupta R (2013) Rice response to site-specific nutrient management in the Indo-Gangetic Plains. Better Crops 97(1):24–26
Jat ML et al (2020) Conservation agriculture and precision nutrient management practices in rice-based systems: impacts on productivity, profitability and environmental sustainability. Agric Syst 180:102792
Jat RA, Wani SP, Sahrawat KL (2022) Integrated nutrient management for improving soil health and crop productivity. Agric Res 11(2):101–114
Kannan S (2010) Foliar fertilization for sustainable crop production. Sustain Agric Rev 4:371–402
Kaur R, Sharma R, Kumar P (2020) Sustainable agriculture policies and practices: a review. J Agric Environ Ethics 33(1):87–107
Kumar N, Kumar R, Singh V (2021) Research priorities for sustainable nutrient management in rice-based systems: a review. Int J Agric Sustain 19(2):124–142
Kumar P, Mishra AK, Chaudhari SK, Singh R, Yadav K, Rai P, Sharma DK (2022a) Conservation agriculture influences crop yield, soil carbon content and nutrient availability in the rice–wheat system of north-west India. Soil Res 60(6):624–635
Kumar P, Fagodiya RK, Chaudhari SK, Singh R, Mishra AK, Singh K, Sharma DK (2022b) Effect of different nitrogen management options on nutrient uptake, biomass carbon sequestration and grain yield of maize-wheat system in reclaimed sodic soil. J Plant Nutr 45(8):1240–1252
Lal R (2021) Organic manures and composts: their role in Integrated Nutrient Management. Int J Environ Sci 3(1):56–72
Lal R et al (2021a) Regenerative agriculture: integrating CA and RA practices to enhance soil health, carbon sequestration, and nutrient cycling in rice-based systems. Soil Tillage Res 205:104759
Lal R, Reicosky D, Hanson JD (2021b) Integration of organic amendments with site-specific nutrient management to enhance soil quality and productivity. J Soil Water Conserv 76(1):11–23
Li Y, Meng F, Wang X, Zhou X, Su B, Zhang X (2018) An intelligent fertilization recommendation system based on the 4R principles for rice production in China. Comput Electron Agric 151:1–13
Liu F, Wang L, Zhang Q (2023a) Importance of iron for improving chlorophyll content and photosynthetic capacity in rice. Plant Sci 325:111–123
Liu S, Huang W, Feng J (2023b) Potassium application increases rice yield and enhances resistance to lodging and pests. Field Crop Res 250:107–116
Luo Y, Su Z, Zhang Y (2021) Enhancing soil health through Integrated Nutrient Management. Soil Biol Biochem 152:108061
Ma BL, Dwyer LM, Costa C, Cober ER, Morrison MJ, Smith DL (1996) Use of physiological leaf traits to diagnose nitrogen status in corn. Can J Plant Sci 76(4):723–730
Miao Y, Stewart BA, Zhang F (2011) Long-term experiments for sustainable nutrient management in China: a review. Agron Sustain Dev 31:397–414. https://doi.org/10.1007/s13593-011-0014-1
Mishra AK, Choudhary PK, Prakash P (2020a) Research priorities for nutrient management in rice cultivation: an Indian perspective. J Crop Improv 34(5):663–684
Mishra AK, Padbhushan R, Biswa LS, Sharma S (2020b) Climate Smart Approach for Water Management in Summer Paddy under Different Establishment Methods in Odisha. National Seminar on Climate Smart Agriculture for Enhancing Farm Profitability (28–29 January, 2020)
Mishra AK, Roohi MS, Grover D, Mishra J (2021) Precise nutrient management in agroecosystems and associated environmental challenges. In: Contemporary environmental challenges India. Rudra Publishers & Distributors, Delhi, pp 203–241
Mowo JG, Kwesiga FR, Berhe T (1993) Nutrient Management in Rice Cultivation. Crop and Weed Science Journal. Retrieved from https://cropandweed.com
Nelson DW, Sommers LE (1996) "Total carbon, organic carbon, and organic matter". Methods of Soil Analysis. Part 3. Chemical Methods. Soil Science Society of America

Pan M, Liu M, Zhang J (2020) A review of integrated models and decision support systems for precision fertilization in China. J Integr Agric 19(9):2245–2257
Parihar CM, Jat HS, Jat SL, Kakraliya SK, Nayak HS (2020) Precision nutrient management for higher nutrient use efficiency and farm profitability in irrigated cereal based cropping systems. Indian J Fertil 16:1000–1014
Pasuquin JM, Castillo EG, Witt C (2004) Leaf colour chart for managing nitrogen fertilization of rice. Better Crops Int 18(1):6–10
Pathak H, Rao DLN, Aggarwal PK (2020) Integrated Nutrient Management for sustainable crop production. Indian J Fertil 16(5):510–519
Peñuelas J, Filella I, Biel C, Serrano L, Savé R (1994) The reflectance at the 950–970 nm region as an indicator of plant water status. Int J Remote Sens 15(7):1235–1247
Phongpan S, Mosier AR, Wassmann R (2000) Mitigating methane emissions from irrigated rice: field evaluation of a water management alternative. Agric Ecosyst Environ 79(3):155–163
Raun WR, Solie JB, Johnson GV, Stone ML, Lukina EV, Thomason WE, Freeman KW (2001) In-season prediction of potential grain yield in winter wheat using canopy reflectance. Agron J 93(1):131–138
Roy A, Chatterjee S, Das P (2020) Soil fertility management through Integrated Nutrient Management. J Plant Nutr 43(12):1771–1784
Sanyal SK, Majumdar K, Singh VK (2014) Nutrient management in Indian agriculture with special reference to nutrient mining - A relook. J Indian Soc Soil Sci 62:307–325
Schmidt JP, Isbell TA, Ferguson RB, DeJong-Hughes J (2021) Site-specific nutrient management: managing variability for improved nutrient use efficiency. Agron J 113(4):2782–2794
Sharma S, Padbhushan R, Kumar U (2019) Integrated nutrient management in rice-wheat cropping system: an evidence on sustainability in the Indian subcontinent through meta-analysis. Agronomy 9:71. https://doi.org/10.3390/agronomy9020071
Sharma A, Thakur P, Kumar V (2020) Role of crop residues in Integrated Nutrient Management. Agric Rev 41(3):202–212
Sharma V, Kumar R, Singh M (2023) Zinc fertilization improves rice yield and grain nutritional quality. J Agron Crop Sci 209(2):237–249
Shoji S, Delgado J, Mosier A, Miura Y (2001) Use of controlled release fertilizers and nitrification inhibitors to increase nitrogen use efficiency and to conserve air and water quality. Commun Soil Sci Plant Anal 32(7–8):1051–1070
Shrivastava AK (2019) Rice fertilization strategies in Chhattisgarh. Journal of Crop Improvement. Retrieved from https://sciencedirect.com
Singh AK, Reddy PP (2020) Use of Azolla in sustainable rice production. Curr Sci 118(2):217–218
Singh U, Ladha JK, Castillo EG, Punzalan GC, Tirol-Padre A, Duqueza M, Watanabe I (2017) Enhancing nitrogen-use efficiency through improved fertilizer management in irrigated rice in the tropics. Field Crop Res 47(2–3):111–120
Singh B, Kumar S, Goel G (2019a) Policies and regulations for sustainable agriculture: a global perspective. Environ Sci Pollut Res 26(26):26652–26668
Singh RP, Yadav MS, Rai M (2019b) Biofertilizers in sustainable agriculture: a review. Int J Agric Biol 21(5):947–956
Singh RK, Kumar P, Sharma V (2020) Nutrient Management in Rice. The Pharma Innovation Journal. Retrieved from https://thepharmajournal.com
Singh VK, Kumar V, Sharma RK (2021) Methodological approaches for site-specific nutrient management in rice: impacts on productivity and sustainability. J Plant Nutr Soil Sci 184(5):738–750
Smith P, Martino D, Cai Z, Gwary D, Janzen H, Kumar P, McCarl B, Ogle S, O'Mara F, Rice C, Scholes B, Sirotenko O, Smith J (2007) Agriculture. In: Metz B, Davidson OR, Bosch PR, Dave R, Meyer LA (eds) Climate Change 2007: Mitigation of Climate Change. Contribution of Working Group III to the Fourth Assessment Report of the Intergovernmental Panel on Climate Change. Cambridge University Press, Cambridge, pp 497–540
Sui R, Zhang W, Li Y (2021) The role of precision agriculture technologies in sustainable agriculture: a review. Precis Agric 22(5):1134–1151

Thapa R, Chatterjee A, Awale R, Daigh A (2016) Deep placement of urea in lowland rice: impact on nitrogen use efficiency, greenhouse gas emissions, and yield. Agron J 108(5):2021–2030
Vance ED, Brookes PC, Jenkinson DS (1987) An extraction method for measuring soil microbial biomass C. Soil Biol Biochem
Viscarra Rossel RA, Behrens T, Ben-Dor E, Brown DJ, Demattê JAM, Shepherd KD, Shi Z, Stenberg B, Stevens A, Adamchuk V, Aïchi H, Barthès BG, Bartholomeus H, Bayer AD, Bernoux M, Brewer TR, Brus DJ, Cardil A, Chappell A, Fouad Y, Genot V, Gomez C, Grunwald S, Gubler A, Guerrero C, Hedley CB, Knadel M, Morrás HJM, Nocita M, Ramirez-Lopez L, Roudier P, Campos EMR, Sanborn P, Sellitto VM, Sudduth KA, Rawlins BG, Walter C, Winowiecki LA, Hong SY, Ji W (2016) A global spectral library to characterize the world's soil. Earth Sci Rev 155:198–230
Wang J, Liu X, Zhang Q et al (2022) Optimizing nitrogen application enhances rice yield and nitrogen use efficiency. J Agric Sci 158(1):45–57
Wang H, Liu J, Xu B (2023a) Manganese application enhances disease resistance and enzyme activities in rice plants. J Plant Nutr 46(3):456–467
Wang L, Wang S, Liu F, Chen X, Zhang F, Cui Z (2023b) Improving rice nitrogen use efficiency and yield through site-specific nutrient management in China. Agron J 115(1):101–113
Witt C, Dobermann A, Abdulrachman S, Gines HC, Wang GH (2005) Chlorophyll meter-based nitrogen management for rice in Asia. Better Crops Int 19(1):10–14
Xu Y, Li H, Zhao J (2022) Calcium improves rice root architecture and increases resistance to stress conditions. Plant Soil 474(1–2):85–98
Yadav RL, Dwivedi BS, Prasad K, Tomar OK, Shurpali NJ, Pandey PS (2021) Yield trends, and changes in soil organic-C and available NPK in a long-term rice-wheat system under Integrated Nutrient Management in north India. Soil Use Manag 16(4):343–349
Yadvinder-Singh, Bijay-Singh, Ladha JK (2004a) Integrated nutrient management for sustainable rice-wheat cropping systems in Asia. Soil Sci Plant Nutr 50(6):1191–1201
Yadvinder-Singh, Bijay-Singh, Ladha JK (2004b) Integrated management of organic manures, crop residues, green manures, and inorganic fertilizers in rice-wheat cropping systems of the Indo-Gangetic Plains. In: Ladha JK, Hill JE, Duxbury JM, Gupta RK, Buresh RJ (eds) Improving the productivity and sustainability of rice-wheat systems: issues and impacts. American Society of Agronomy and International Rice Research Institute, Madison, WI, pp 197–217
Yang XM, Li XL, Zhang FS (2003) Crop nitrogen status assessment using leaf color charts in rice and wheat. Commun Curr Res Educ Top Trends Appl Microbiol 1:369–376
Yang Y, Liu Q, Hu Z (2023) Copper application enhances antioxidant enzyme activities and mitigates oxidative stress in rice. Environ Exp Bot
Zhang W, Xu A, Liu B, Pan X (2014) The impact of water-saving irrigation and controlled-release nitrogen fertilizer on yield, fertilizer-N use efficiency and greenhouse gas emissions of paddy rice. Field Crop Res 155:193–201
Zhao Y, Wang X, Li M (2023) Sulfur application improves rice growth and grain protein content. Crop Sci 63(2):589–601
Zhu Y, Chen L, Zhao X (2023) Phosphorus fertilization improves rice grain quality and stress resistance. Plant Nutr Soil Sci 185(3):235–248

Pathways for Assessment and Managing Soil Biodiversity

8

Debadatta Sethi, Shraddha Mohanty, Kshitipati Padhan, Sanjib Kumar Sahoo, Prava Kiran Dash, Konathala Kusumavathi, Kiran Kumar Mohapatra, and Narayan Panda

D. Sethi
Department of Soil Science and Agricultural Chemistry, Odisha University of Agriculture and Technology, Bhubaneswar, India

AICRP on PHET, College of Agricultural Engineering and Technology, Odisha University of Agriculture and Technology, Bhubaneswar, India

S. Mohanty
Krishi Vigyan Kendra, Odisha University of Agriculture and Technology, Kalahandi, India

K. Padhan
Department of Agriculture and Allied Science, CV Raman Global University, Bhubaneswar, India

S. K. Sahoo
College of Agriculture, Odisha University of Agriculture and Technology, Bhawanipatna, India

P. K. Dash
Regional Research and Technology Transfer Station, Odisha University of Agriculture and Technology, Dhenkanal, India

K. Kusumavathi
Department of Agronomy, Bidhan Chandra Krishi Viswavidyalaya, Mohanpur, Nadia, West Bengal, India

Indian Institute of Water Management, ICAR, Bhubaneswar, India

K. K. Mohapatra
AICRP for Dryland Agriculture, Odisha University of Agriculture and Technology, Phulbani, India

N. Panda (✉)
Department of Soil Science and Agricultural Chemistry, Odisha University of Agriculture and Technology, Bhubaneswar, India

A. K. Mishra et al. (eds.), *Key Drivers and Indicators of Soil Health Management*, https://doi.org/10.1007/978-981-97-7564-4_8

Introduction

The soil, as one of Earth's most intricate habitats, supports an extraordinary diversity of life. Approximately 25–30% of all species on the planet are believed to inhabit soils either throughout their entire life or during specific stages (Tibbett et al. 2020). Soil organisms play a crucial role in ecosystem services by aiding livestock production and crop, antibiotics and pathogen hosting, nutrient regulation in both water and soil, and overseeing the circulation of greenhouse gases. Additionally, they contribute to biomass generation, influencing climate change (Bardgett and van der Putten 2014; Adhikari and Hartemink 2016; Chen et al. 2020). Global distribution of soil biodiversity appears to deviate from the patterns observed in aboveground plants and animals. Unlike aboveground organisms, which often display higher diversity in tropical regions along latitudinal gradients, there is limited evidence supporting a similar pattern in soil biodiversity (Bardgett and van der Putten 2014). Thorough examinations of global soil biodiversity unveil a notable degree of diversity, with each soil presenting a unique species composition. These compositions are either confined to specific geographical areas or marked by a significant level of endemism (Fierer et al. 2012). Rotifers possess a phylogenetically acquired ability to utilize conglomerates and aggregates as a form of "food" (Datki et al. 2021).

Nevertheless, soil depth distinctly impacts the multiplicity, uniformity, and microbial diversity. Soil depth can modify the rates of microbial evolution, influencing factors such as community congregation and the overall structure of the microbial diversity. A reduction in microbial community richness and diversity became evident as soil depth increased (Hao et al. 2021) (Fig. 8.1). Generally, subsurface soil microbial communities tend to be relatively simple, contrasting with the high complexity observed in surface soil microbial diversity (Du et al. 2018; Xin et al. 2023). The content and diversity of bacterial and archaeal communities are impacted by the depth of the soil (Feng et al. 2019). However, molecular studies suggest that environmental factors such as climate and soil conditions, including pH, play a more significant role in shaping microbial communities than their mobility (Fierer et al. 2012). In general, elevated levels of belowground biodiversity may not align with the "hotspots" of aboveground biodiversity, as seen in regions like the tropics (Fierer et al. 2012).

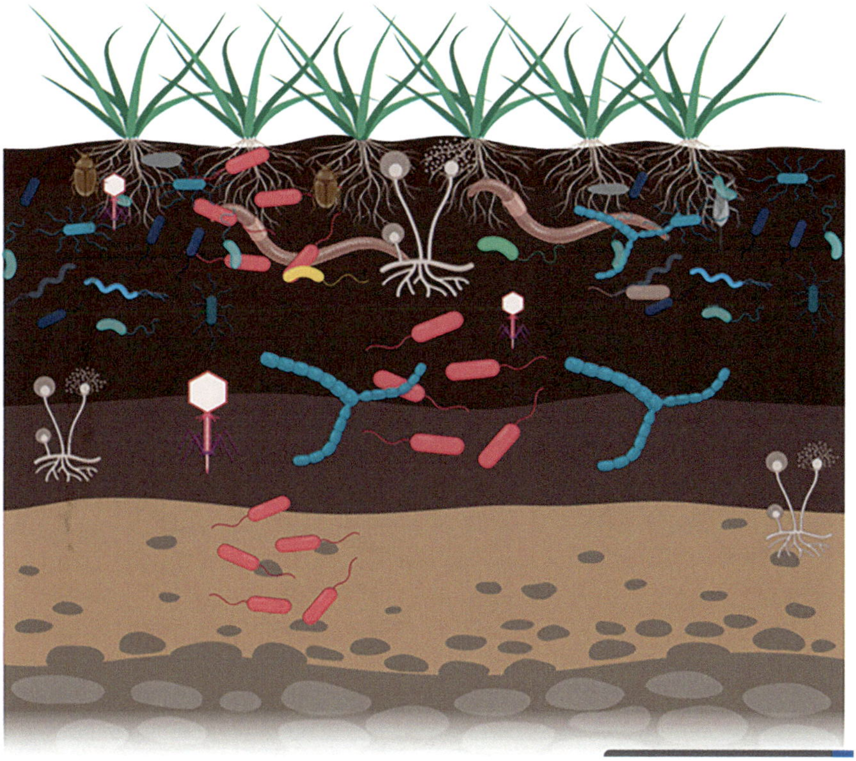

Fig. 8.1 Different types of organisms beneath the soil surface, and its diversity along the depth

Threats to Soil Biodiversity

The decline in soil biodiversity is commonly marked by a reduction in both the quantity and variety of life-forms inhabiting soils (Jones et al. 2005). In regions experiencing a decline in soil biodiversity, there can be significant repercussions for the normal functioning of the soil, its responsiveness to disturbances, and its ability to recover. Several threats to soil, including intensive human exploitation, changes in land use, and decline of soil organic matter, have been identified as factors negatively impacting soil biodiversity (Gardi et al. 2013). The correlation between soil biodiversity and other soil degradation processes has been recognized (Turbé et al. 2010). A total of 12 potential threats were identified, with the order of importance as per experts' weighting in accordance with Jeffery and Gardi (2010). The recognized threats to soil biodiversity (Fig. 8.2) include human intensive exploitation, soil organic matter decline, sealing, compaction, erosion, habitat fragmentation (Gardi et al. 2013; Orgiazzi et al. 2015), habitat disruption, soil pollution, land-use change, climate change, invasive species, and GMO pollution (Aksoy et al. 2017). These threats have been evaluated to gauge their magnitude, and endeavors have

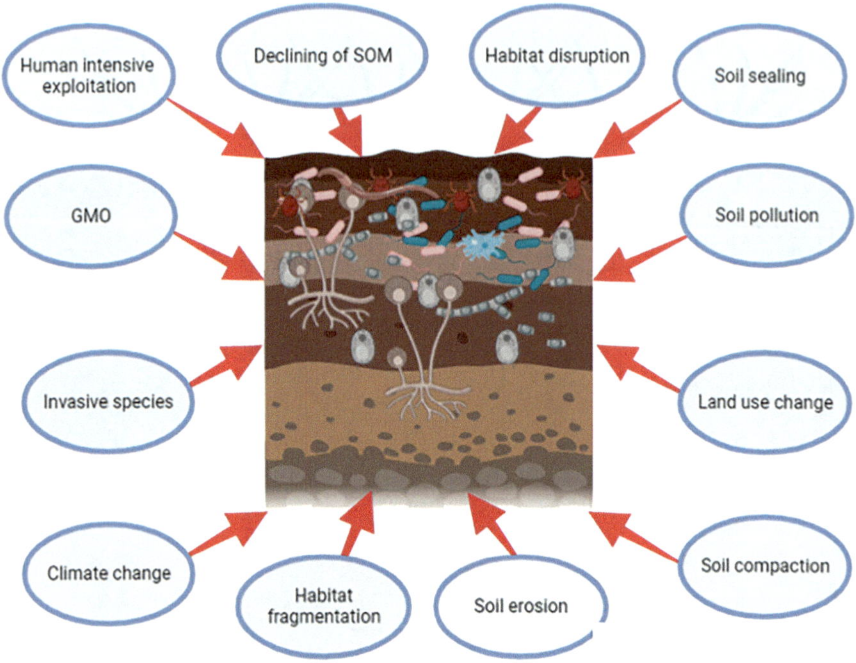

Fig. 8.2 Different threats to soil biodiversity

been made to map their potential impacts on soil biodiversity (Jeffery and Gardi 2010).

The response to identified threats in both biologically active soil and normal soil exhibited similar patterns of interference (Fig. 8.3). However, in normal soil, the process of becoming 100% biologically functional took a considerable amount of time. Following disturbances, the soils initially demonstrated resistance for a limited duration, after which a breakdown occurred. Upon reaching a resistant point again, there was an attempt to recover, showcasing its inherent ability to regain its original status, a characteristic known as resilience. In the case of normal soil, the initial microbial growth was gradual, followed by an exponential increase, eventually reaching a plateau stage.

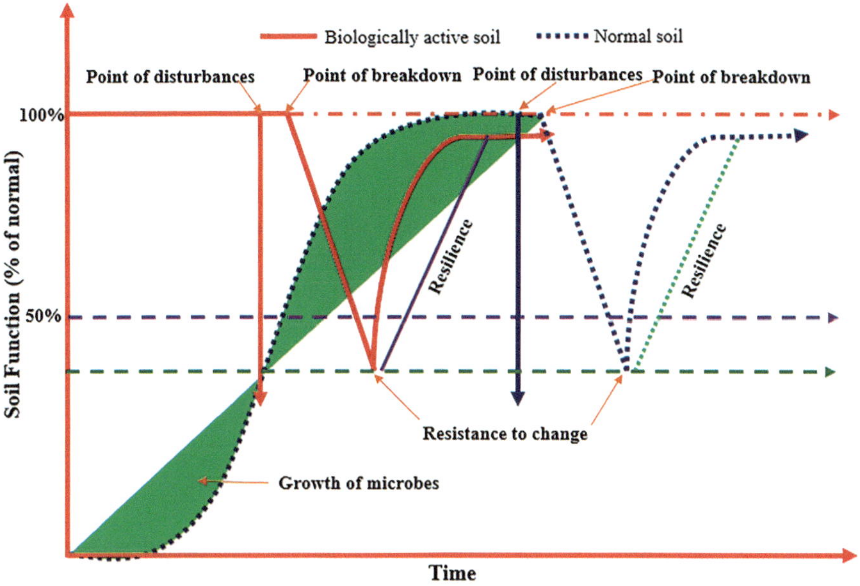

Fig. 8.3 Comparison between a normal and biologically active soil with respect to resistance and resilience of a soil biological function

Different Strategies for Assessment of Soil Biodiversity

Soil biodiversity assessment is crucial for understanding the soil health and nutrient-supplying and recycling capacity of the soil ecosystem involving different biogeochemical and ecological processes. The soil microbial diversities are studied using different technologies, from conventional to cutting-edge molecular approaches. Many conventional technologies have been employed to evaluate soil biodiversity, enabling fundamental understandings of the microbial dynamics. Some of them are discussed below.

Conventional Technologies of Soil Biodiversity Assessment

Several conventional technologies have been adopted for assessing the functions of soil biodiversity, viz. microscopic approaches, biochemical methods, plate count method, sole carbon source utilization (SCSU), and fatty acid methyl ester (FAME) approach. There are some drawbacks in conventional approaches like these techniques are unable to differentiate between closely related taxa. Biochemical methods are used to quantify metabolic activities (Blagodatskaya and Kuzyakov 2013). Plate count technique is inherently biased towards cultivable microbes in the laboratory (Stevenson et al. 2004). SCSU approach assesses metabolic diversity of microbial communities (Preston-Mafham et al. 2002). Biolog microplates (Garland and

Mills 1991) are commonly used for screening of microbial communities. This technique is also used for community-level physiological profiling of the microbial diversity (Wang et al. 2006). Fatty acid methyl ester (FAME) approach is an alternative or supplementary method to study microbial diversity particularly in environmental samples like soil, water, and sediments (Lacey et al. 2021). This approach furnishes details about the composition of microbial communities by analyzing the various fatty acids present in their cell walls (Soni and Devi 2022). As a result, the presence or varying ratios of specific fatty acid methyl esters (FAMEs) can serve as indicators for the presence and quantitative abundance of particular taxa in a sample, essentially functioning as biomarkers. For instance, the neutral lipid 16:1ω5 is commonly used as an indicator of arbuscular mycorrhizal fungi in soils. In the case of *Phytophthora*, the addition of *P. sojae* zoospores to soils led to elevated levels of 18:2ω6 to 22:1ω6 (Lacey et al. 2021).

Modern Technologies of Soil Biodiversity Assessment

The biodiversity of soil is a vast subterranean ecosystem that includes an array of diverse species, including archaea, insects, nematodes, fungi, bacteria, and earthworms. However, heterogeneity of soil is a barrier in producing a large-scale framework of the relationship between soil and the organism living within it (Aksoy et al. 2017).

Extracellular Environmental DNA (eDNA)

The advent of DNA-based species identification, utilizing extensive reference databases of standardized DNA barcodes such as COI, rbcL-matK, and ITS, has enabled the development of high-throughput techniques for assessing the composition of communities from bulk and environmental samples (Creer et al. 2016). Environmental eDNA-based biodiversity assessment methodologies involve identifying species in an ecosystem by extracting DNA directly from environmental substrates like water, soil, and air, without isolating the target organisms (Taberlet et al. 2012). Importantly, repeat sampling is unnecessary, as the same eDNA sample can be used to identify different species groups (Dysthe et al. 2018). This approach has transformed ecological community examination by leveraging various sources of DNA in an environmental sample (Deiner et al. 2017). The eDNA extracted from samples aims to provide a representative snapshot of the total biodiversity in the sampled location. Individual RNA sequences, often referred to as operational taxonomical units (OTUs) or amplicon sequence variants (ASVs), represent different species (Becker and Pushkareva 2023).

Metagenomics

Previously microbial biodiversity's significance was gauged through limited processes like denitrification, primary production, or degradation of organic carbon (Pedler et al. 2014; Hou et al. 2015; Jung et al. 2016). The metagenome approach is employed to elucidate the fundamental microbiota in various domains, including crops, soil, gut microbiome, and wastewater (Rajendhran and Gunasekaran 2008;

Singh et al. 2022). Metagenomics combines conventional microbiology and molecular biology to study all the species in an environmental sample simultaneously. Metagenomics is a culture-independent technique employed to assess microbial diversity in soil samples (Rivera and Izard 2015; Hardwick et al. 2018). It identifies the range of uncultivated soil microbial species through the direct isolation and analysis of genetic materials (DNA) from specific habitats. The extensive sequence information derived from environmental DNA can be employed not only to investigate biodiversity but also to analyze member functions within the community (Becker and Pushkareva 2023).

Assessment of Soil Biodiversity Through Mapping

An assessment is made possible by the present understanding of how soil physico-chemical characteristics affect biota. The connection and interactions between soil biodiversity and external factors are elucidated below. Temperature and precipitation rates are altered by the rising CO_2 concentrations in the atmosphere brought on by climate change, and these changes collectively affect the availability of soil organic matter (Aksoy et al. 2017). Land use, cover, and management practices significantly influence soil biodiversity, with grassland soils typically exhibiting higher biodiversity compared to forests, farmed areas, or urban land (Turbé et al. 2010). Intense monocropping promotes the establishment of only a portion of soil microorganisms, which results in a decrease in biodiversity, whereas naturally occurring diversified vegetation increases soil biodiversity (Figuerola et al. 2015). The datasets for various variables mentioned earlier have been reclassified based on the thresholds provided in Table 8.1, and the results are visually represented in the

Table 8.1 Different variables with their threshold value and related ratting which might have a strong effect on soil biodiversity

Sl. No.	Variable	Threshold value and related scores					
1	pH	Classes	<4	4–5.2	5.2–8.2	>8.2	
		Rating	1	2	3	2	
2	Textural classes	Classes	1	2	3	4	5
		Rating	1	2	2	2	2
3	Organic matter (%)	Classes	<1	1–2	2–4	>4	
		Rating	1	2	3	4	
4	P-ETP (mm)	Classes	<500	500	>500		
		Rating	1	2	3		
5	Annual average temperature (°)	Classes	<5	5–20	>20		
		Rating	1	2	1		
6	Soil biomass productivity	Classes	Poor	Average	Good		
		Rating	1	2	3		
7	Land use/land cover	Classes	Artificial	Arable	Permanent crops	Others	
		Rating	0	1	1.5	2	

Source: Aksoy et al. (2017)

P-ETP: potential evapotranspiration; textural class codes: 1. coarse; 2. medium; 3. medium-fine; 4. fine; 5. very fine maps. Rutgers et al. (2016) studied and published a map of earthworm communities in Europe, and another work by Griffiths et al. (2016) used geostatistical techniques to map and predict soil bacterial biodiversity utilizing datasets at the European and national levels. There is a lack of integrated approach as most of the studies focused only on the single aspect of soil biodiversity, hence the need to extend towards a multifactorial approach.

Different Strategies for Management of Soil Biodiversity

There is a mounting body of evidence indicating that the combined effects of elevated organic matter inputs and reduced tillage work synergistically enhance biological activity in soil, potentially improving resilience to extreme events.

Amendments and Inputs

Soil physicochemical and biological properties are all impacted by the various management practices, which in turn affect the populations of soil microbes (Chojnacka et al. 2019). Soil additives might be organic or inorganic (Ulzen et al. 2018).

Organic: Organic additives obtained from living sources increase microbial activity and produce organic acids and other metabolites that reduce nutrient sorption, increase nutrient availability, and revive microbiological characteristics (Hernandez et al. 2015). Out of these, the most frequently used amendment in agriculture is animal manures. Management of agro-residues through composting (Patra et al. 2022; Mishra et al. 2022a; Padhan et al. 2023) and vermicomposting (Pandit et al. 2020; Sethi et al. 2023; Padhan et al. 2024) could be able to enhance soil biodiversity utilized as a potting medium or soil conditioner (Barthod et al. 2018). However, urban wastes, particularly organic municipal solid waste, sewage, sludge, and industry waste, have limited use in crop field because of the presence of heavy metals and other contaminants (Petrie et al. 2014). On-farm composting recycles agricultural biomasses reliably, cheaply, and sustainably (Pane et al. 2015). Biochar is another common organic amendment used in agriculture (Pati and Panda 2021; Nogués et al. 2023). Biochar quality depends on its substrate as well as pyrolysis temperature (Bolan et al. 2021). It improves soil biological activity (Liang et al. 2006) and water-holding capacity (Diatta et al. 2020) and reduces fertilizer dosages and leaching (Sulok et al. 2021). Microbial inoculants and biofertilizers improve crop production by increasing nutrient availability, nutrient uptake, and water-use efficiency (Sethi et al. 2019a, b; Subudhi et al. 2020; Sethi et al. 2021; Athul et al. 2022; Verma et al. 2022). Manure is stabilized by composting, which reduces pathogens (Borrero et al. 2004) and creates a safe soil addition (Franco et al. 2006).

Green manures refer to incorporation of plant biomass into the soil. Green manuring can enhance physicochemical properties of soil, augment the organic matter concentration, enhance the activities of soil microbes, and ultimately lead to increased crop production (Krobel et al. 2014; Patil et al. 2014; Kusumavathi et al.

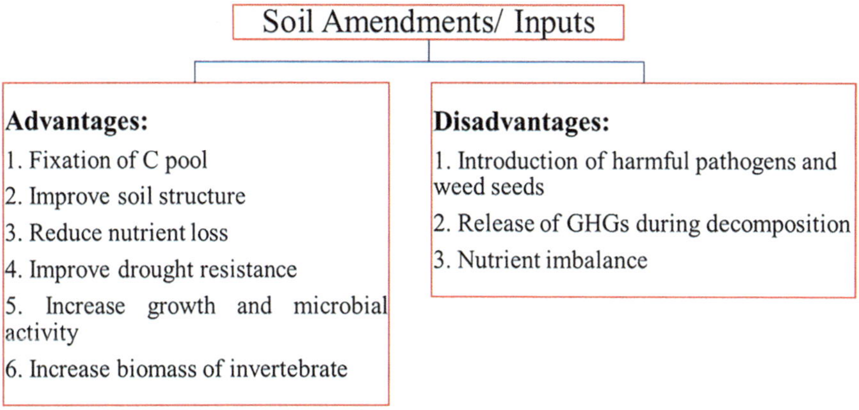

Fig. 8.4 Advantage and disadvantage of soil amendments

2018; Mangaraj et al. 2023). The inoculation of well-characterized microbial strains or their combinations (Singhai et al. 2011) enhances the soil quality. In scientific literature, these types of populations are called "synthetic communities" (Mazzola and Freilich 2017). Lehmann et al. (2020) and Ninh et al. (2015) found that native bacteria can be enhanced by including organic substances (Athul et al. 2022), in contrast to synthetic communities. Many types of microorganisms may be affected by the use of microbial amendments; however, an integrated approach may have more benefits and could be useful (Ray et al. 2020).

Inorganic: Inorganic additives are used in soil reclamation and site-specific management strategies. This strategy is implemented in a sensible and environmentally suitable manner, ultimately leading to the enrichment of soil biodiversity (Urra et al. 2019) and enhancing physicochemical and biological characteristics of soil (Hernandez et al. 2014; Chocano et al. 2016; Readyhough et al. 2021). Incorporating gypsum into soil enhances its conditions in several ways, including root extension, nutrient and water utilization, subsurface Al toxicity reduction, and good yield response (Zoca and Penn 2017). According to Baghbani-Arani et al. (2020), use of zeolite gives significant result because it improves soil physical properties, increases cation-exchange capacity, and decreases nitrogen leaching (Fig. 8.4).

Cultural Practices

In general, it is seen that the conventional method of tillage, monocropping, and injudicious use of agrochemicals are major factors which have a negative influence on soil biodiversity. Reduced tillage practices resulted in higher moisture, organic matter, microbial diversity, and enzymatic activity over conventional tillage (Gajda et al. 2018). Soil biodiversity is positively influenced by crop diversification (Zhang et al. 2020). Crop diversification is frequently characterized as the deliberate introduction of diversity into cropping systems. This practice is not only vital for maximizing crop production but also plays a crucial role in improving soil health. It achieves this by maintaining a balance in soil biodiversity, improving soil nutrient-use efficiency, and mitigating the impact of soilborne pathogens (Gurr et al. 2016; Barbieri et al. 2019; Yang et al. 2020). Therefore, adopting intercropping, mixed cropping, frequent crop rotations, and agroforestry will facilitate the natural cycling of nutrients and other inputs, thus favoring preserving of biodiversity (Campbell et al. 1991; Mishra et al. 2022b). Keeping the soil covered with crop residues acts as mulch, which prevents weed infestation, allows moisture conservation, and protects soils from erosion by wind and water. Including a buffer zone of tree species within and around the farm improves both above- and belowground species population. Terrace farming with grassed waterways is a good practice in hilly terrains. Application of broad-spectrum herbicide and insecticides greatly hampers natural soil biodiversity. Therefore, managing insect pests by encouraging more numbers of natural enemies will preserve the soil biological population.

Conclusion

Numerous and varied threats pose challenges to soil biodiversity, and the current body of literature has only recently begun to unravel the complexities of soil biological systems. Our knowledge of the entities within these systems is limited, and a comprehensive understanding of their functional roles, niche partitioning, and interactions among these organisms is yet to be fully grasped. We have to inculcate different modern techniques like metagenomics for accessing biological diversity in the soil and different strategies like organic farming, natural farming, bioinoculation, and reclamation of problem soil in a judicious way for enhancing functional soil biodiversity.

References

Adhikari K, Hartemink AE (2016) Linking soils to ecosystem services - a global review. Geoderma 262:101–111. https://doi.org/10.1016/j.geoderma.2015.08.009

Aksoy E, Louwagie G, Gardi C, Gregor M, Schröder C, Löhnertz M (2017) Assessing soil biodiversity potentials in Europe. Sci Total Environ 589:236–249

Athul PP, Patra RK, Sethi D, Panda N, Mukhi SK, Padhan K, Sahoo SK, Sahoo TR, Mangaraj S, Pradhan SR, Pattanayak SK (2022) Efficient native strains of rhizobia improved nodulation and productivity of French bean (Phaseolus vulgaris L.) under rainfed condition. Front Plant Sci 13:1048696. https://doi.org/10.3389/fpls.2022.1048696

Baghbani-Arani A, Jami MG, Namdari A, Borz-Abad RK (2020) Influence of irrigation regimes, zeolite, inorganic and organic manures on water use efficiency, soil fertility and yield of sunflower in a sandy soil. Commun Soil Sci Plant Anal 51:711–725

Barbieri P, Pellerin S, Seufert V, Nesme T (2019) Changes in crop rotations would impact food production in an organically farmed world. Nat Sustain 2:378–385

Bardgett R, van der Putten W (2014) Belowground biodiversity and ecosystem functioning. Nature 515:505–511. https://doi.org/10.1038/nature13855

Barthod J, Rumpel C, Dignac MF (2018) Composting with additives to improve organic amendments. A review. Agron Sustain Dev 38:17. https://doi.org/10.1007/s13593-018-0491-9

Becker B, Pushkareva E (2023) Metagenomics provides a deeper assessment of the diversity of bacterial communities in polar soils than metabarcoding. Genes 14:812. https://doi.org/10.3390/genes14040812

Blagodatskaya E, Kuzyakov Y (2013) Active microorganisms in soil: critical review of estimation criteria and approaches. Soil Biol Biochem 67:192–211

Bolan N, Hoang SA, Beiyuan J, Gupta S, Hou D, Karakoti A, Joseph S, Jung S, Kim KH, Kirkham MB, Kua HW, Kumar M, Kwon EE, Ok YS, Perera V, Rinklebe J, Shaheen SM, Sarkar B, Sarmah AK, Singh BP, Singh G, Tsang DCW, Vikrant K, Vithanage M, Vinu A, Wang H, Wijesekara H, Yan Y, Younis SA, Van Zwieten L (2021) Multifunctional applications of biochar beyond carbon storage. Int Mater Rev. https://doi.org/10.1080/09506608.2021.1922047

Borrero C, Trillas MI, Ordovás J, Tello JC, Avilés M (2004) Predictive factors for the suppression of Fusarium wilt of tomato in plant growth media. Phytopathology 94:1094–1101

Campbell C, Biederbeck V, Zentner R, Lafond G (1991) Effect of crop rotations and cultural practices on soil organic matter, microbial biomass and respiration in a thin Black Chernozem. Can J Soil Sci 71:363–376

Chen XD, Dunfield KE, Fraser TD, Wakelin SA, Richardson AE, Condron LM (2020) Soil biodiversity and biogeochemical function in managed ecosystems. Soil Res 58:1–20. https://doi.org/10.1071/SR19067

Chocano C, Garcia C, González D, Melgares de Aguilar J, Hernandez T (2016) Organic plum cultivation in the Mediterranean region: the medium-term effect of five different organic soil management practices on crop production and microbiological soil quality. Agric Ecosyst Environ 221:60–70

Chojnacka K, Gorazda K, Witek-Krowiak A, Moustakas K (2019) Recovery of fertilizer nutrients from materials - Contradictions, mistakes and future trends. Renew Sust Energ Rev 110:485–498

Creer S, Deiner K, Frey S, Porazinska D, Taberlet P, Thomas WK, Potter C, Bik HM (2016) The ecologist's field guide to sequence-based identification of biodiversity. Methods Ecol Evol 7:1008–1018. https://doi.org/10.1111/2041-210X.12574

Datki Z, Acs E, Balazs E, Sovany T, Csoka I, Zsuga K, Kalman J, Galik-Olah Z (2021) Exogenic production of bioactive filamentous biopolymer by monogonant rotifers. Ecotoxicol Environ Saf 208:111666. https://doi.org/10.1016/j.ecoenv.2020.111666

Deiner K, Bik HM, Mächler E, Seymour M, Lacoursière-Roussel A, Altermatt F, Creer S, Bista I, Lodge DM, de Vere N, Pfrender ME, Bernatchez L (2017) Environmental DNA metabarcoding: transforming how we survey animal and plant communities. Mol Ecol 26:5872–5895. https://doi.org/10.1111/mec.14350

Diatta A, Fike JH, Battaglia ML, Galbraith JM, Baig MB (2020) Effects of biochar on soil fertility and crop productivity in arid regions: a review. Arab J Geosci 13:595. https://doi.org/10.1007/s12517-020-05586-2

Du C, Xu C-Y, Jian J-S, He W-X, Hou L, Geng Z-C (2018) Seasonal dynamics of bacterial communities in a Betula albosinensis forest. Eur J Soil Sci. https://doi.org/10.1111/ejss.12568

Dysthe JC, Rodgers T, Franklin TW, Carim KJ, Young MK, McKelvey KS, Mock KE, Schwartz MK (2018) Repurposing environmental DNA samples—Detecting the western pearlshell (Margaritifera falcata) as a proof of concept. Ecol Evol 8(5):2659–2670. https://doi.org/10.1002/ece3.3898

Feng H, Guo J, Wang W, Song X, Yu S (2019) Soil depth determines the composition and diversity of bacterial and archaeal communities in a poplar plantation. Forests 10:550. https://doi.org/10.3390/f10070550

Fierer N, Leff JW, Adams BJ, Nielsen UN, Bates ST, Lauber CL, Owens S, Gilbert JA, Wall DH, Caporaso JG (2012) Cross-biome metagenomic analyses of soil microbial communities and their functional attributes. Proc Natl Acad Sci U S A 109(52):21390–21395

Figuerola EL, Guerrero LD, Türkowsky D, Wall LG, Erijman L (2015) Crop monoculture rather than agriculture reduces the spatial turnover of soil bacterial communities at a regional scale. Environ Microbiol 17(2):678–688

Franco A, Schuhmacher M, Roca E, Domingo JL (2006) Application of cattle manure as fertilizer in pastureland: Estimating the incremental risk due to metal accumulation employing a multi-compartment model. Environ Int 32(6):724–732

Gajda AM, Czyż EA, Dexter AR, Furtak KM, Grządziel J, Stanek-Tarkowska J (2018) Effects of different soil management practices on soil properties and microbial diversity. Int Agrophys 32(1):81

Gardi C, Jeffery S, Saltelli A (2013) An estimate of potential threats levels to soil biodiversity in the EU. Glob Chang Biol 19(5):1538–1548. https://doi.org/10.1111/gcb.1215

Garland JL, Mills AL (1991) Classification and characterization of heterotrophic microbial communities on the basis of patterns of community-level sole-carbon-source utilization. Appl Environ Microbiol 57(8):2351–2359

Griffiths RI, Thomson BC, Plassart P, Gweon HS, Stone D, Creamer RE, Lemanceau P, Bailey MJ (2016) Mapping and validating predictions of soil bacterial biodiversity using European and national scale datasets. Appl Soil Ecol 97:61–68

Gurr GM, Lu Z, Zheng X, Xu H, Zhu P, Chen G, Yao X, Cheng J, Zhu Z, Catindig JL, Villareal S, Van Chien H, Cuong LQ, Channoo C, Chengwattana N, Lan LP, Hai LH, Chaiwong J, Nicol HI, Perovic DJ, Wratten SD, Heong KL (2016) Multi-country evidence that crop diversification promotes ecological intensification of agriculture. Nat Plants 2:16014

Hao J, Chai YN, Lopes LD, Ordóñez RA, Wright EE, Archontoulis S, Schachtman DP (2021) The effects of soil depth on the structure of microbial communities in agricultural soils in Iowa (United States). Appl Environ Microbiol 87(8):e02673–e02620. https://doi.org/10.1128/AEM.02673-20

Hardwick SA, Chen WY, Wong T, Kanakamedala BS, Deveson IW, Ongley SE, Santini NS, Marcellin E, Smith MA, Nielsen LK, Lovelock CE, Neilan BA, Mercer TR (2018) Synthetic microbe communities provide internal reference standards for metagenome sequencing and analysis. Nat Commun 9:3096

Hernandez T, Chocano C, Moreno JL, Garcia C (2014) Towards a more sustainable fertilization: Combined use of compost and inorganic fertilization for tomato cultivation. Agric Ecosyst Environ 196:178–184

Hernandez T, Garcia E, Garcia C (2015) A strategy for marginal semiarid degraded soil restoration: a sole addition of compost at a high rate. A five-year field experiment. Soil Biol Biochem 89:61–71

Hou L, Yin G, Liu M, Zhou J, Zheng Y, Gao J, Zong H, Yang Y, Gao L, Tong C (2015) Effects of sulfamethazine on denitrification and the associated N2O release in estuarine and coastal sediments. Environ Sci Technol 49(1):326–333

Jeffery S, Gardi C (2010) Soil biodiversity under threat—a review. Acta Soc Zool Bohem 74:7–12

Jones RJA, Hiederer R, Rusco E, Montanarella L (2005) Estimating organic carbon in the soils of Europe for policy support. Eur J Soil Sci 56(5):655–671. https://doi.org/10.1111/j.1365-2389.2005.00728.x

Jung J, Philippot L, Park W (2016) Metagenomic and functional analyses of the consequences of reduction of bacterial diversity on soil functions and bioremediation in diesel-contaminated microcosms. Sci Rep 6:23012. https://doi.org/10.1038/srep23012

Krobel R, Lemke R, Campbell CA, Zentner R, McConkey B, Steppuhn H, De Jong R, Wang H (2014) Water use efficiency of spring wheat in the semi-arid Canadian prairies: effect of legume green manure, type of spring wheat, and cropping frequency. Can J Soil Sci 94:223–235

Kusumavathi K, Pattanayak SK, Mohapatra AK, Sethi D (2018) Effect of nutrient management on soil fertility in rice (Oryza sativa)-greengram (Vigna radiata) cropping system. Ann Plant Soil Res 20:330–337

Lacey RF, Sullivan-Hill BA, Deslippe JR, Keyzers RA, Gerth ML (2021) The Fatty Acid Methyl Ester (FAME) profile of Phytophthora agathidicida and its potential use as diagnostic tool. FEMS Microbiol Lett 368(17): fnab113. https://doi.org/10.1093/femsle/fnab113

Lehmann J, Bossio DA, Kögel-Knabner I, Rillig MC (2020) The concept and future prospects of soil health. Nat Rev Earth Environ 1:544–553. https://doi.org/10.1038/s43017-020-0080-8

Liang B, Lehmann J, Solomon D, Kinyangi J, Grossman J, O'Neill B, Skjemstad JO, Thies J, Luizao FJ, Petersen J, Neves EG (2006) Black carbon increases cation exchange capacity in soils. Soil Sci Soc Am J 70:1719–1730

Mangaraj S, Paikaray RK, Garnayak LM, Behera SD, Patra B, Sethi D, Pradhan SR, Jena R (2023) Crop and soil productivity of a rice-green gram system under integrative nutrient management. Agron J 115:2631–2645. https://doi.org/10.1002/agj2.21401

Mazzola M, Freilich S (2017) Prospects for biological soilborne disease control: application of indigenous versus synthetic microbiomes. Phytopathology 107:256–263. https://doi.org/10.1094/PHYTO-09-16-0330-RVW

Mishra AK, Grover D, Das R, Mishra J, Biswal B (2022a) Carbon sequestration potential in agricultural systems. In: Agro-biodiversity and agri-ecosystem management. Springer, Singapore, pp 87–111

Mishra AK, Sharma S, Chattopadhyay P, Grover D, Chattopadhyay P, Mishra J (2022b) Natural farming system as hotspots of ecotourism and their potential for sustainable development. In: Strategies to achieve sustainable development (Sdgs): a roadmap for global development. Nova Science Publishers, New York, pp 189–206

Ninh HT, Grandy AS, Wickings K, Snapp SS, Kirk W, Hao J (2015) Organic amendment effects on potato productivity and quality are related to soil microbial activity. Plant Soil 386:223–236. https://doi.org/10.1007/s11104-014-2223-5

Nogués I, Miritana VM, Passatore L, Zacchini M, Peruzzi E, Carloni S, Pietrini F, Marabottini R, Chiti T, Massaccesi L, Marinari S (2023) Biochar soil amendment as carbon farming practice in a Mediterranean environment. Geoderma Reg 33:e00634. https://doi.org/10.1016/j.geodrs.2023.e00634

Orgiazzi A, Dunbar MB, Panagos P, de Groot GA, Lemanceau P (2015) Soil biodiversity and DNA barcodes: Opportunities and challenges. Soil Biol Biochem 80:244–250

Padhan K, Patra RK, Sethi D, Panda N, Sahoo S, Pattanayak SK, Senapati AK (2023) Isolation, characterization, and identification of cellulose-degrading bacteria for composting of agro-wastes. Biomass Convers Biorefin. https://doi.org/10.1007/s13399-023-04087-y

Padhan K, Patra RK, Sethi D, Mohanty S, Sahoo SK, Panda N, Pattanayak SK, Patra AK (2024) Exploitation of cellulose-degrading bacteria in bioconversion of agro-wastes. Chemosphere 347:140654. https://doi.org/10.1016/j.chemosphere.2023.140654

Pandit L, Sethi D, Pattanayak SK, Nayak Y (2020) Bioconversion of lignocellulosic organic wastes into nutrient-rich vermicompost by Eudrilus eugeniae. Bioresour Technol Rep 12:100580. https://doi.org/10.1016/j.biteb.2020.100580

Pane C, Celano G, Piccolo A, Villecco D, Spaccini R, Palese AM, Zaccardelli M (2015) Effects of on-farm composted tomato residues on soil biological activity and yields in a tomato cropping system. Chem Biol Technol Agric 2:1–13

Pati MM, Panda N (2021) Effect of sources of biochar on soil biological properties in an acidic Alfisols. Ann Plant Soil Res 23:523–525

Patil GD, Kharade MS, Sawale DD (2014) Effect of green manures on physico-chemical properties of soil. J Agric Res Technol 39:160–163

Patra RK, Behera D, Mohapatra KK, Sethi D, Mandal M, Patra AK, Balasubramani R (2022) Juxtaposing the quality of compost and vermicompost produced from organic waste amended with cow dung. Environ Res 214:114119. https://doi.org/10.1016/j.envres.2022.114119

Pedler BE, Aluwihare LI, Azam F (2014) Single bacterial strain capable of significant contribution to carbon cycling in the surface ocean. Proc Natl Acad Sci U S A. 111:7202–7207

Petrie B, Barden R, Kasprzyk-Hordern BA (2014) Review on emerging contaminants in wastewaters and the environment: current knowledge, understudied areas and recommendations for future monitoring. Water Res 72:3–27

Preston-Mafham J, Boddy L, Randerson PF (2002) Analysis of microbial community functional diversity using sole-carbon-source utilisation profiles–a critique. FEMS Microbiol Ecol 42:1–14

Rajendhran J, Gunasekaran P (2008) Strategies for accessing soil metagenome for desired applications. Biotechnol Adv 26:576–590

Ray P, Lakshmanan V, Labbé JL, Craven KD (2020) Microbe to microbiome: a paradigm shift in the application of microorganisms for sustainable agriculture. Front Microbiol 11:1–15. https://doi.org/10.3389/fmicb.2020.622926

Readyhough T, Neher D, Andrews T (2021) Organic amendments alter soil hydrology and belowground microbiome of tomato (Solanum lycopersicum). Microorganisms 9:1561

Rivera MC, Izard J (eds) (2015) Promises and prospects of microbiome studies, 1st edn. Amsterdam, Elsevier

Rutgers M, Orgiazzi A, Gardi C, Römbke J, Jänsch S, Keith AM, Neilson R, Boag B, Schmidt O, Murchie AK, Blackshaw RP (2016) Mapping earthworm communities in Europe. Appl Soil Ecol 97:98–111

Sethi D, Mohanty S, Pattanayak SK (2019a) Acid and salt tolerance behavior of Rhizobium isolates and their effect on microbial diversity in the rhizosphere of redgram (Cajanus cajan). Indian J Biochem Biophys 56:245–252

Sethi D, Mohanty S, Pattanayak SK (2019b) Effect of different carbon, nitrogen and vitamin sources on exopolysaccharide production of Rhizobium species isolated from root nodule of redgram. Indian J Biochem Biophys 56:86–93

Sethi D, Subudhi S, Rajput VD, Kusumavathi K, Sahoo TR, Dash S, Mangaraj S, Nayak DK, Pattanayak SK, Minkina T, Glinushkin AP, Kalinitchenko VP (2021) Exploring the role of mycorrhizal and rhizobium inoculation with organic and inorganic fertilizers on the nutrient uptake and growth of Acacia mangium saplings in acidic soil. Forests 12:1657. https://doi.org/10.3390/f12121657

Sethi D, Kusumavathi K, Ravindran B, Panda N, Padhan K, Dash S, Sahoo TR, Mangaraj S, Dhal A, Swain SK, Sarkar S, Pattanayak SK, Febrisiantosa A (2023) Bioconversion of organic wastes into wealth by vermitechnology: a review. In: Ravindran B, Gupta SK, Bhat SA, Chauhan PS, Tyagi N (eds) Recent trends in solid waste management. Woodhead Publishing-Elsevier, Cambridge, pp 27–53. https://doi.org/10.1016/B978-0-443-15206-1.00009-8

Singh N, Singh V, Rai SN, Vamanu E, Singh MP (2022) Metagenomic analysis of garden soil-derived microbial consortia and unveiling their metabolic potential in mitigating toxic hexavalent chromium. Life 12:2094. https://doi.org/10.3390/life12122094

Singhai PK, Sarma BK, Srivastava JS (2011) Biological management of common scab of potato through Pseudomonas species and vermicompost. Biol Control 57:150–157. https://doi.org/10.1016/j.biocontrol.2011.02.008

Soni R, Devi S (2022) Composting process: fundamental and molecular aspects (Chapter 12). In: Samuel J, Kumar A, Singh J (eds) Relationship between microbes and the environment for sustainable ecosystem services, volume 1: microbial products for sustainable ecosystem services. Elsevier. https://doi.org/10.1016/C2020-0-02873-2

Stevenson BS, Eichorst SA, Wertz JT, Schmidt TM, Breznak JA (2004) New strategies for cultivation and detection of previously uncultured microbes. Appl Environ Microbiol 70:4748–4755

Subudhi S, Sethi D, Pattanayak SK (2020) Characterization of Rhizobium sp. (SAR-5) isolated from root nodule of Acacia mangium L. Indian J Biochem Biophys 57:327–333

Sulok KMT, Ahmed OH, Khew CY, Zehnder JAM, Jalloh MB, Musah AA, Abdu A (2021) Chemical and biological characteristics of organic amendments produced from selected agrowastes with potential for sustaining soil health: a laboratory assessment. Sustainability 13:4919

Taberlet P, Coissac E, Hajibabaei M, Rieseberg LH (2012) Environmental DNA. Mol Ecol 21:1789–1793. https://doi.org/10.1111/j.1365-294X.2012.05542.x

Tibbett M, Fraser TD, Duddigan S (2020) Identifying potential threats to soil biodiversity. PeerJ 8:e9271. https://doi.org/10.7717/peerj.9271

Turbé A, De Toni A, Benito P, Lavelle P, Camacho NR, van Der Putten WH, Labouze E, Mudgal S (2010) Soil Biodiversity: Functions, Threats and Tools for Policy Makers. Bio Intelligence Service, IRD, and NIOO, Report for European Commission. DG Environment, Brussels, Belgium

Ulzen J, Abaidoo RC, Ewusi-Mensah N, Masso C (2018) On-farm evaluation and determination of sources of variability of soybean response to Bradyrhizobium inoculation and phosphorus fertilizer in northern Ghana. Agric Ecosyst Environ 267:23–32

Urra A, Wadham J, Hawkings JR, Telling J, Hatton JE, Yde JC, Hasholt B, van As D, Bhatia MP, Nienow P (2019) Weathering dynamics under contrasting Greenland ice sheet catchments. Front Earth Sci 7:299

Verma H, Patra RK, Sethi D, Pattanayak SK (2022) Isolation and characterization of native Rhizobium from root nodules of Raikiafrench bean growing area of Odisha. Indian J Biochem Biophys. 59:918–926. https://doi.org/10.56042/ijbb.v59i9.61519

Wang MC, Gong M, Zang HB, Hua XM, Yao J, Pang YJ, Yang YH (2006) Effect of methamidophos and urea application on microbial communities in soils as determined by microbial biomass and community level physiological profiles. J Environ Sci Health B 41:399–413

Xin Y, Wu Y, Zhang H, Li X, Qu X (2023) Soil depth exerts a stronger impact on microbial communities and the sulfur biological cycle than salinity in salinized soils. Sci Total Environ 894:164898. https://doi.org/10.1016/j.scitotenv.2023.164898

Yang T, Siddique KHM, Liu K (2020) Cropping systems in agriculture and their impact on soil health - A review. Global Ecol Conserv 23:e01118. https://doi.org/10.1016/j.gecco.2020.e01118

Zhang J, Van Der Heijden MG, Zhang F, Bender SF (2020) Soil biodiversity and crop diversification are vital components of healthy soils and agricultural sustainability. Front Agric Sci Eng 7:236

Zoca SM, Penn C (2017) An important tool with no instruction manual: a review of gypsum use in agriculture. Adv Agron 144:1–44

Efficient Water Management Strategies and Recharge Potential Under Rice-Based Cropping Systems

9

Girija Prasad Patnaik, Ajay Kumar Mishra, Sheetal Sharma, and Tarini Prasad Das

Introduction

Rice is a staple food for over half of the world's population and is predominantly grown in water-intensive paddy fields (Mallareddy et al. 2023). Effective water management in rice-based cropping systems is critical not only for ensuring sustainable rice production but also for conserving water resources and enhancing groundwater recharge (Bouman et al. 2007a, b; Bhatt et al. 2021). This chapter delves into various water management strategies that can be employed in rice cultivation, examining their impacts on water-use efficiency and groundwater recharge potential. Rice-based cropping systems are diverse agricultural practices primarily reliant on flooded paddy fields, varying significantly based on climatic, soil, and water availability conditions. Traditional flooded systems maintain continuous flooding throughout most of the growing season, which helps control weeds and pests and preserves soil nutrients but demands a constant and ample water supply, posing challenges in water-scarce areas. In contrast, the alternate wetting and drying (AWD) system allows fields to dry intermittently before re-flooding, conserving water, and reducing methane emissions, though it requires precise water management and timing to ensure optimal plant growth. Aerobic rice systems, where rice is grown in non-flooded, well-drained soils, are more water efficient and similar to cultivating upland crops but face increased weed pressure and necessitate efficient irrigation techniques. Declining groundwater table due to its over- and injudicious use in agriculture mainly in water-intensive crops and cropping system like rice has raised environmental conservation issues which need to be solved. Innovative rice

G. P. Patnaik (✉) · A. K. Mishra · S. Sharma
International Rice Research Institute, New Delhi, India

T. P. Das
Department of Agronomy, Odisha University of Agriculture and Technology, Bhubaneswar, Odisha, India

A. K. Mishra et al. (eds.), *Key Drivers and Indicators of Soil Health Management*, https://doi.org/10.1007/978-981-97-7564-4_9

production system such as direct seeded rice has the potential to recharge groundwater with higher infiltration and percolation through pervious layer of soil. However, this chapter captures insights regarding the efficient water management practices and different ways to recharge groundwater in rice-based cropping systems.

Challenges in Water Management

Effective water management is a critical aspect of rice-based cropping systems, facing numerous challenges that necessitate the adoption of more efficient practices to ensure sustainability and productivity.

Water Scarcity

Water scarcity has become an increasingly pressing issue due to rising populations and growing agricultural demands. Traditional rice cultivation is particularly water intensive (Bouman et al. 2007a, b), often requiring large volumes of water to maintain flooded conditions necessary for optimal growth. This high water demand places significant pressure on available water resources, especially in regions already experiencing water shortages. As global water resources become more strained, it is imperative for farmers and agricultural stakeholders to adopt water-saving practices and technologies. Strategies such as alternate wetting and drying (AWD), direct seeded rice (DSR), system of rice intensification (SRI), use of drought-resistant rice varieties, and improved irrigation infrastructure can help reduce water consumption and mitigate the impacts of water scarcity on rice production (Waris et al. 2019; Vijayakumar et al. 2023).

Environmental Concerns

The excessive use of water in rice fields can lead to several environmental issues that have both local and global implications. One major concern is the depletion of local water bodies (Bouman 2007; Chapagain and Hoekstra 2011). When large volumes of water are diverted for rice cultivation, it can lead to reduced water levels in rivers, lakes, and aquifers, adversely affecting local ecosystems and biodiversity. Additionally, continuous flooding of rice fields contributes significantly to methane emissions (Neue 1993; Cai et al. 2003), a potent greenhouse gas that exacerbates climate change. Another environmental issue linked to excessive water use is soil salinization (Nachshon 2018). In areas with poor drainage or high evaporation rates, the accumulation of salts in the soil can occur, reducing soil fertility and adversely affecting crop yields. This salinization problem can render large areas of agricultural land unproductive if not managed properly. Furthermore, the intensive use of water for rice fields often reduces the availability of water for downstream users.

This can impact other agricultural activities, domestic water supply, and industrial uses, leading to conflicts over water resources and affecting the overall water security of a region. To address these environmental concerns, adopting more sustainable water management practices is crucial (Russo et al. 2014). Techniques such as precision irrigation, maintaining field water levels that minimize unnecessary water loss, and integrating water-saving technologies can help reduce the environmental footprint of rice cultivation.

Efficient Water Management Strategies

Alternate Wetting and Drying (AWD)

It is an innovative and efficient water-saving technology designed to improve the sustainability of rice cultivation. This method involves allowing the rice field to dry intermittently during the growing period, rather than maintaining continuous flooding. The practice of AWD is carefully managed by allowing the field to dry out to a certain depth (typically 15 cm below the soil surface) before re-flooding it (Lampayan et al. 2015; Norton et al. 2017). This cycle of wetting and drying is repeated throughout the growing season, providing several significant benefits. One of the primary advantages of AWD is its potential to significantly reduce water usage. By allowing the fields to dry periodically, AWD can lower water consumption by up to 30% compared to traditional continuous flooding methods (Lampayan et al. 2015; Sumana et al. 2019; Ishfaq et al. 2020a, b). This reduction is crucial in areas facing water scarcity, helping to conserve valuable water resources and ensuring their availability for other agricultural and domestic needs. Additionally, the decreased water use can lead to cost savings for farmers, who may face high expenses related to irrigation. Another key benefit of AWD is the reduction in methane emissions (Setyanto et al. 2018; Cheng et al. 2022). Continuous flooding of rice fields creates anaerobic conditions that favour the production of methane, a potent greenhouse gas. By intermittently drying the fields, AWD introduces aerobic conditions that inhibit methane production, thereby lowering overall emissions. This environmental benefit is particularly important in the context of climate change mitigation, as it helps to reduce the carbon footprint of rice production. AWD also positively impacts the health and growth of rice plants. The drying periods encourage stronger root development as the roots are forced to grow deeper in search of moisture. This improved root growth enhances the plants' ability to absorb nutrients and water, leading to healthier and more resilient crops (Zhang et al. 2009; Santiago-Arenas et al. 2019). Moreover, the intermittent drying helps to reduce the incidence of certain waterborne diseases and pests, which can thrive in continuously flooded conditions. This can result in reduced reliance on chemical pesticides, further contributing to the environmental sustainability of rice farming. In addition to these benefits, AWD can improve the overall efficiency of water management in rice fields. Farmers adopting AWD often use simple tools such as field water tubes to monitor water levels accurately and decide the optimal timing for irrigation. This

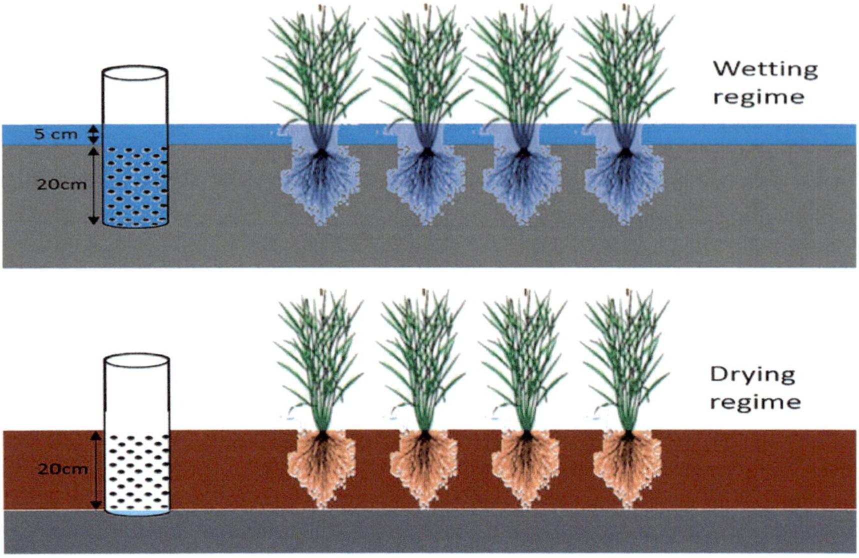

Fig. 9.1 Alternate wetting and drying (AWD) irrigation of rice (Adapted from Riaz et al. 2018)

precision in water management not only conserves water but also ensures that the rice plants receive adequate moisture at critical growth stages (Fig. 9.1).

System of Rice Intensification (SRI)

The system of rice intensification (SRI) is an innovative agricultural methodology aimed at increasing the productivity of rice cultivation while reducing the use of water and other inputs (Satyanarayana et al. 2007; Thakur et al. 2014). SRI involves a set of practices that differ significantly from traditional rice farming methods. Key components of SRI include planting fewer seedlings with wider spacing, using organic inputs, and managing water more efficiently. These practices collectively contribute to a more sustainable and productive rice farming system, offering several notable benefits. One of the primary advantages of SRI is the enhancement of water-use efficiency (Reddy et al. 2005; Zhao et al. 2009). Traditional rice cultivation often involves continuous flooding of fields, leading to substantial water consumption. In contrast, SRI promotes the intermittent irrigation of fields, which helps maintain soil moisture without the need for constant flooding. This method conserves water by reducing evaporation and percolation losses, making it particularly beneficial in regions facing water scarcity (Choudhury et al. 2007). By optimizing water use, SRI not only conserves a vital resource but also reduces the costs associated with irrigation, offering economic benefits to farmers. Another critical benefit of SRI is the promotion of vigorous root development. In SRI, seedlings are transplanted at a younger age, typically 8–12 days old, compared to the older seedlings used in traditional methods (Uphoff and Randriamiharisoa 2002; Uphoff 2003; Patra and Haque 2011). Additionally, the seedlings are planted individually rather than in clumps, with wider spacing between plants. The increased spacing and younger transplanting age encourage the development of a more robust root system. Deeper and more extensive roots enhance the plant's ability to access nutrients and water from the soil, leading to healthier and more resilient rice plants capable of withstanding environmental stresses such as drought and pests. SRI also leads to increased yields, a crucial factor for farmers looking to maximize their production. The combination of improved root health, better nutrient uptake, and reduced plant competition under SRI practices results in higher tillering, more panicles, and a greater number of grains per panicle. These factors collectively contribute to a significant increase in overall yield. Research and field trials have consistently shown that SRI can lead to yield improvements ranging from 15% to 45% or more (Satyanarayana et al. 2007; Vijayakumar et al. 2006; Omwenga et al. 2014), depending on the local conditions and the extent of practice adoption. This shift towards organic inputs improves soil health over time by enhancing soil structure, increasing microbial activity, and reducing the build-up of harmful chemicals. Healthier soils contribute to more sustainable farming practices, as they are better able to retain moisture and nutrients, leading to continued high productivity without degrading the natural resource base. SRI is a transformative agricultural practice that significantly enhances water conservation in rice cultivation. Unlike traditional methods that rely on continuous flooding, SRI employs intermittent irrigation and shallow water management, drastically reducing water use. By avoiding constant flooding, SRI minimizes water loss through evaporation and seepage. Besides, techniques such as transplanting young seedlings at wider spacing and employing precision irrigation contribute to further water savings. Studies have shown that SRI can reduce water usage by 20–50% compared to conventional rice farming methods (Vijayakumar et al. 2006; Satyanarayana et al. 2007; Thakur et al. 2014; Omwenga et al. 2014; Abd El-Latif and Abdullahb 2018).

Direct Seeded Rice (DSR)

Direct seeded rice (DSR) is an alternative rice cultivation method where seeds are sown directly into the field, bypassing the traditional practice of transplanting seedlings (Pathak et al. 2011; Farooq et al. 2011). This approach offers several advantages, particularly in terms of water saving and water-use efficiency, which are critical in the context of increasing water scarcity and the need for sustainable agricultural practices.

One of the most significant benefits of DSR is its potential to reduce water use substantially. In traditional rice farming, fields are typically flooded to a depth of several centimetres for most of the growing season. This continuous flooding not only consumes vast amounts of water but also contributes to water wastage through evaporation, seepage, and percolation. In contrast, DSR eliminates the need for such prolonged flooding. By sowing seeds directly into the field, DSR allows for more precise water management. This method can lead to water savings of up to 30–50% compared to traditional flooded rice systems (Bouman et al. 2007a, b; Talebnejad and Sepaskhah 2014; Hang et al. 2022). Such a significant reduction in water use is particularly beneficial in areas facing water shortages or where water resources are limited. DSR also shortens the cropping duration, which further enhances water-use efficiency. In traditional transplanting, the process of raising seedlings in nurseries and then transplanting them into flooded fields can be time consuming, often extending the overall cropping period. DSR eliminates the nursery stage and the associated transplantation shock, allowing rice plants to establish and grow faster (Kumar and Ladha 2011). The shorter cropping duration means that the rice plants require irrigation for a reduced period, thereby conserving water (Singh and Bhattacharyya 1989). Traditional rice transplanting is labour intensive, requiring significant manual effort to transplant seedlings from nurseries to the main field. This labour-intensive process often involves maintaining flooded conditions to facilitate easy transplantation. Farmers can use broadcasting techniques or seed drills to automate the sowing process and lessen their dependency on manual labour by implementing DSR (Joshi et al. 2011; Kaur and Singh 2017). In addition to lowering labour costs, the automating of sowing operations and the decreased requirement for continual flooding help to make better use of available water resources. Furthermore, the direct seeding approach in DSR encourages the development of a more extensive root system. When rice seeds are sown directly into the field, the plants can establish deeper root systems compared to transplanted seedlings. Deeper roots allow rice plants to access water and nutrients more effectively from the soil, enhancing their resilience to water stress and reducing the frequency of irrigation needed (Panda et al. 2021; Liu et al. 2022). This improved root architecture contributes to better water-use efficiency, as the plants are better able to sustain themselves with less frequent watering. Direct seeded rice (DSR) offers a range of benefits that enhance water saving and water-use efficiency by eliminating the need for continuous flooding, shortening the cropping duration, and reducing labour costs (Table 9.1).

Aerobic Rice System

Aerobic rice represents a significant shift from traditional flooded rice cultivation, involving the growth of rice in non-flooded, well-drained soils (Prasad 2011). This method is designed to address various challenges associated with water conservation, labour reduction, and soil health enhancement. One of the most notable benefits of aerobic rice systems is the substantial reduction in water consumption. Aerobic rice, on the other hand, is cultivated under non-flooded conditions, similar to other upland crops like wheat or maize unlike conventional puddled transplanted rice (Prasad 2011; Nie et al. 2012). By avoiding continuous flooding, aerobic rice systems drastically cut down the amount of water needed for irrigation (Peng et al. 2006). This water-saving aspect is particularly critical in regions where water scarcity is a growing concern, as it allows for more sustainable use of limited water resources (Tuong and Bouman 2003). Because there is less water required,

Table 9.1 Different resource conservation rice production systems

Sl. No.	Rice production system	Water saving upon continuous flooding	Remarks	Reference
1.	Alternate wetting and drying	20–50%	Reduction in the greenhouse gas emission and yield advantage upon continuous flooding	Kumar and Rajitha (2019)
		25–70%	Reduction in CH_4 emission up to 95%	Ishfaq et al. (2020a, b)
		20–60%	AWD boosted grain yield by 15% during the wet season and 7% during the dry season when compared to continuous flooding. Additionally, water productivity increased by 46–77% relative to continuous flooding	Sumana et al. (2019)
		38%	Alternate wetting and drying does not reduce yields and profits of rice farmers	Lampayan et al. (2015)
		16.9–21.9% Upon continuous flooding	Increase in yield due to increased nutrient-use efficiency	Boruah et al. (2018)
		11.88% Upon continuous flooding	Reduced CH_4 emission by 23.10% but increased N_2O emission by 14.79%	Sriphirom et al. (2019)
		19.5% Upon continuous flooding	AWD practices' yield increased compared to farmer practices	Faruki et al. (2011)
		24% and 38% upon continuous flooding	AWD had higher water productivity than continuously flooded rice	Yao et al. (2012)
2.	Direct seeded rice	49–55% Water saving	Higher yields (13–18%) and increase in water productivity (22–30%)	Ishfaq et al. (2020a, b)
		38–47% Water saving	Reduced GHG emission, increased water productivity, and sustained rice yield and 88.9% higher water productivity	Hang et al. (2022)
		15–23% Water saving	Water-use efficiency (WUE) was higher with DSR system. Direct seeding of rice on raised beds	Bhushan et al. (2007)
		53% Reduction in irrigation water use	Yield and straw and root dry matter (DM) similar to conventional	Talebnejad and Sepaskhah (2014)

(continued)

Table 9.1 (continued)

Sl. No.	Rice production system	Water saving upon continuous flooding	Remarks	Reference
3.	System of rice intensification	40% Water saving	5–10% Yield increase over conventional method	Satyanarayana et al. (2007)
		32.4% Water saving	46.4% Yield increase over conventional method	Omwenga et al. (2014)
		22% Water saving	9.1% Yield advantage over conventional method	Vijayakumar et al. (2006)
		25% Water saving	25% and 50% increase in yields	Katambara et al. (2013)
		28–35% Water saving	20.9% Yield advantage	Abd El-Latif and Abdullahb (2018)
		18–21% Water saving	20–30% Yield advantage over conventional flooding	Biswas et al. (2021)
		28.8% Water saving	SRI produced 49% higher grain yield compared to conventional flooding	Thakur et al. (2014)

smallholder farmers may find rice production to be more financially feasible and irrigation expenditures can be decreased. The fact that aerobic rice systems require less labour is another important benefit. Aerobic rice can be sown directly into the field, eliminating the need for transplanting unlike puddled transplanted rice. This direct seeding approach, often facilitated by mechanization, reduces the manual labour involved in rice cultivation. Moreover, managing non-flooded fields is generally less laborious than maintaining continuously flooded paddies, freeing up time and resources for farmers to engage in other agricultural activities or pursue additional income-generating opportunities. Aerobic rice systems also contribute to enhanced soil health. Flooded rice fields are prone to various soil health issues, such as reduced microbial diversity and soil structure degradation due to prolonged water saturation. In aerobic systems, well-drained soils promote better aeration, which supports a more diverse and active soil microbial community unlike conventional rice system. Improved soil aeration and structure also enhance root growth, allowing rice plants to access nutrients and water more efficiently (Kumar et al. 2018).

Groundwater Recharge Potential

Rice (*Oryza sativa* L.) is a crucial crop and a staple food for nearly half of the global population, providing over 20% of the calories consumed by humans worldwide (Smith 1998). It accounts for 40% of all global irrigation and 17% of global groundwater depletion. The global average water footprint per kilogram of rice is 2500 L, while in India, it is 2800 L (Bouman 2009). Rice can be grown using traditional transplanting methods or through direct seeded rice (DSR), where seeds are sown directly into the soil, bypassing the need to transplant seedlings from a nursery. The DSR method saves both labour and time (Mann and Ashraf 2004; Jehangir et al. 2007).

Various studies (Yadav et al. 2011; Mohammad et al. 2018) have used the soil-water balance approach to estimate deep percolation, water movement, and surface runoff in transplanted and direct-seeded rice fields. Researchers have also assessed the potential for groundwater recharge from flooded rice fields under different climate change scenarios using the HYDRUS-1D numerical model and MODFLOW (Kambale et al. 2017; Patle et al. 2017). Direct seeding of rice, whether in dry or wet conditions, is an efficient water-saving method that enhances irrigation efficiency and deep-water percolation, potentially contributing to groundwater recharge (Kumar et al. 2015; Choudhary et al. 2016). Effective water management in rice-based systems can significantly influence groundwater recharge.

Percolation and Seepage

Deep percolation and seepage play vital roles in groundwater recharge, particularly in agricultural settings where crop fields are abundant. Understanding their significance requires delving into the intricate relationship between water, soil, and crops and how these processes impact groundwater replenishment. Deep percolation refers to the downward movement of water through the soil profile beyond the root zone of plants (Bethune et al. 2008). This typically occurs when irrigation or rainfall exceeds the soil's capacity to retain water, leading to saturation and subsequent percolation below the root zone. Seepage, on the contrary, involves the lateral movement of water through soil layers (Pandey et al. 2007), often occurring when water infiltrates the soil surface and migrates laterally due to variations in soil permeability.

The significance of deep percolation and seepage in groundwater recharge from crop fields can be understood through several key aspects. Deep percolation and seepage contribute to maintaining the water balance in agricultural systems (LaHue and Linquist 2021). They help prevent surface runoff, which can lead to soil erosion and nutrient loss, by allowing excess water to infiltrate into the soil and potentially recharge groundwater reserves (Seiler and Gat 2007). In regions where groundwater is a significant source of irrigation water, deep percolation and seepage serve as crucial mechanisms for replenishing aquifers. As water moves through the soil profile and laterally through permeable layers, it gradually replenishes groundwater reserves, ensuring their sustainability for future agricultural and non-agricultural needs.

Rice fields are typically flooded to a certain depth to support rice cultivation. As water infiltrates the soil, some of it moves downwards beyond the root zone due to gravity. This process is known as deep percolation. Deep percolation is significant for groundwater recharge because it replenishes aquifers, especially in regions where groundwater levels have depleted due to over-extraction or seasonal variations. In rice cultivation, deep percolation helps to maintain groundwater levels by allowing excess water to seep into the underlying layers, thus recharging the aquifers (Seiler and Gat 2007; Nazarieh et al. 2018).

Deep percolation and seepage play several roles in agricultural ecosystems, encompassing nutrient transport, root zone management, salinity control, and ecosystem services (Li et al. 2023). They facilitate the movement of nutrients, fertilizers, and agrichemicals through the soil, potentially enhancing soil fertility and crop growth if managed effectively. Moreover, efficient management of these processes is crucial for maintaining optimal moisture levels within the root zone, preventing waterlogging, and promoting healthy root development. In regions prone to soil salinity, deep percolation aids in flushing salts below the root zone, mitigating salinity issues, and improving soil quality for cultivation. By replenishing groundwater reserves, deep percolation and seepage support vital ecosystem services, sustaining wetlands, riparian zones, and freshwater habitats essential for wildlife, thus contributing to overall ecosystem health and resilience. They act as natural mechanisms to mitigate the impacts of climate change on water availability while also supporting the adoption of practices that ensure the efficient use and conservation of water resources in rice cultivation and beyond. In rice fields, seepage occurs when water from the flooded fields gradually infiltrates the surrounding soil and eventually percolates into the groundwater system (LaHue and Linquist 2019). Seepage is essential for groundwater recharge as it facilitates the transfer of water from surface reservoirs to underground aquifers. By allowing water to slowly seep into the ground, seepage ensures a continuous replenishment of groundwater resources, which is crucial for sustaining agricultural activities and meeting water demands in both rural and urban areas.

Impact of Tillage and Rice Production System

Different tillage methods in rice cultivation, such as puddled, non-puddled, and direct-seeded rice, have unique effects on soil structure, deep percolation, and groundwater recharge. In puddled rice cultivation, the tillage method involves repetitive ploughing and harrowing of soil under flooded conditions to establish a soft, muddy bed, effectively breaking down soil aggregates and forming a compacted layer known as the plough pan beneath the surface (Sharma and Bhagat 1993). Ultimately, the soil structure becomes less permeable due to the destruction and compaction of aggregates, restricting water movement through the soil. This compacted plough pan plays a pivotal role in reducing deep percolation losses by retaining water in the upper soil layers, thus maintaining the flooded conditions essential for rice cultivation. However, this reduction in deep percolation results in limited water movement to deeper soil layers and the groundwater table (Goitom et al. 2016), leading to minimal groundwater recharge from puddled fields. Conversely, in non-puddled rice cultivation, which entails less intensive soil disturbance through practices like reduced tillage or zero tillage, the soil structure remains more porous with intact aggregates, fostering higher soil permeability and facilitating better water infiltration. Consequently, increased soil permeability in non-puddled systems leads to higher rates of deep percolation, enabling greater water movement to deeper layers and positively contributing to groundwater recharge (Kroes et al. 2019). Similarly, in direct-seeded rice (DSR) cultivation, where seeds are sown directly into the field without transplanting seedlings, both dry and wet DSR methods maintain better soil structure and porosity compared to puddled fields, enhancing permeability and facilitating significant water movement through the soil profile. Consequently, increased deep percolation in DSR practices significantly contributes to groundwater recharge, enhancing the sustainability of groundwater resources.

Impact of Puddling

Research indicates that puddling reduces vertical water movement due to the formation of a compacted plough pan. For instance, Bouman et al. (2007a, b) found that puddled fields had lower deep percolation rates compared to non-puddled fields. The study emphasized that the low permeability of the puddled layer effectively maintains flooded conditions required for rice but limits groundwater recharge.

Impact of Non-puddled and Direct-Seeded Rice

Studies suggest that non-puddled and direct-seeded systems have higher rates of deep percolation and groundwater recharge. Singh et al. (2016) showed that DSR systems have increased water infiltration rates, leading to higher groundwater recharge compared to puddled systems. The study highlighted that maintaining soil structure and porosity in DSR enhances water movement through the soil, contributing to sustainable water management.

Role of Direct-Seeded Rice in Groundwater Recharge

Significant water conservation benefits are attainable through direct-seeded rice (DSR) cultivation (Dawe 2005). Effective water management is essential, particularly during the crop emergence phase, spanning the initial 7–15 days post-sowing in DSR (Balasubramanian and Hill 2002; Kumar et al. 2009). It is imperative to maintain soil moisture at optimal levels without saturation to prevent seed rotting. Following sowing in dry soil, the application of flush irrigation becomes crucial if rainfall is unlikely, while saturating the field at the three-leaf stage is imperative (Bouman et al. 2007a, b). This approach fosters robust rooting and seedling establishment while stimulating weed seed germination (Kamboj et al. 2012). In Northwest India, DSR cultivation in non-puddled soils resulted in water savings ranging from 35% to 57% (Singh et al. 2002). Small-scale trials demonstrated a notable 20% reduction in irrigation requirements (Gupta et al. 2003). Implementing raised-bed planting methods enhanced water management and decreased irrigation demand, albeit with a slight reduction in yield (Balasubramanian and Hill 2002; Gupta et al. 2003; Hossain et al. 2003). DSR cultivation on raised beds notably reduced water usage by 12–60% and concurrently increased yield by 10% compared to transplanted rice (TPR) (Gupta et al. 2003). Moreover, water productivity in DSR surpassed that of TPR, indicative of superior water-use efficiency (Gill et al. 2006). Optimal levelling of lands in DSR cultivation may further enhance water-use efficiency (WUE) and productivity (Joshi et al. 2013). Effective bund management

Table 9.2 The comparison between TPR and DSR with respect to potential groundwater recharge (Adapted from Gulati et al. 2022)

	Estimated value (mm)	Predicted value (mm)
TPR		
Crop evapotranspiration	545	526
Surface run-off	416	442
Potential groundwater recharge	1139.4	1131
DSR		
Crop evapotranspiration	539	506
Surface run-off	152.9	152
Potential groundwater recharge	1305.6	1336

is also critical for regulating water depth and minimizing losses (Lantican et al. 1999; Humphreys et al. 2010). DSR cultivation aids groundwater recharge through increased infiltration and deep percolation of applied water, further contributing to sustainable water management practices. Dry direct seeding is a method with potential to save both water and labour in rice cultivation. It can reduce labour requirements by 50% for crop establishment, with further savings possible if mechanized seeding is used. Dry direct seeding can yield comparable or higher results than puddle-transplanted rice. It also reduces greenhouse gas emissions and build-up of arsenic and other heavy metals and improves soil health compared to conventional systems (Hang et al. 2022).

In an experiment at Punjab, India, Gulati et al. (2022) established a comparison between the potential groundwater recharge of DSR and TPR system. Deep percolation, estimated through the soil-water balance equation, amounted to 1139.4 mm for transplanted rice (TPR) and 1305.6 mm for direct-seeded rice (DSR), as outlined in Table 9.2. DSR showed approximately 14.6% more deep percolation compared to TPR, attributed to its higher infiltration rate in non-puddled and porous root-zone soil. The total water inputs measured for TPR and DSR were 2035.5 mm and 1935.5 mm, respectively. Crop evapotranspiration represented 26.77% and 27.85% of the total water input for TPR and DSR, while run-off accounted for 20.43% and 7.89% for TPR and DSR, respectively. Deep percolation beyond 100 cm depth constituted 55.95% and 67.45% of the total water input for TPR and DSR, respectively (Fig. 9.2).

Seasonal Fluctuation

Seasonal fluctuations play a significant role in influencing aquifer recharge, as variations in precipitation, evapotranspiration, and agricultural water usage occur throughout the year. The wet season, characterized by heavy rainfall, leads to increased infiltration rates, replenishing groundwater levels and boosting aquifer recharge (Tanaka et al. 2010; Wang et al. 2014). This period is crucial for restoring groundwater reservoirs, particularly in areas with distinct dry and wet seasons.

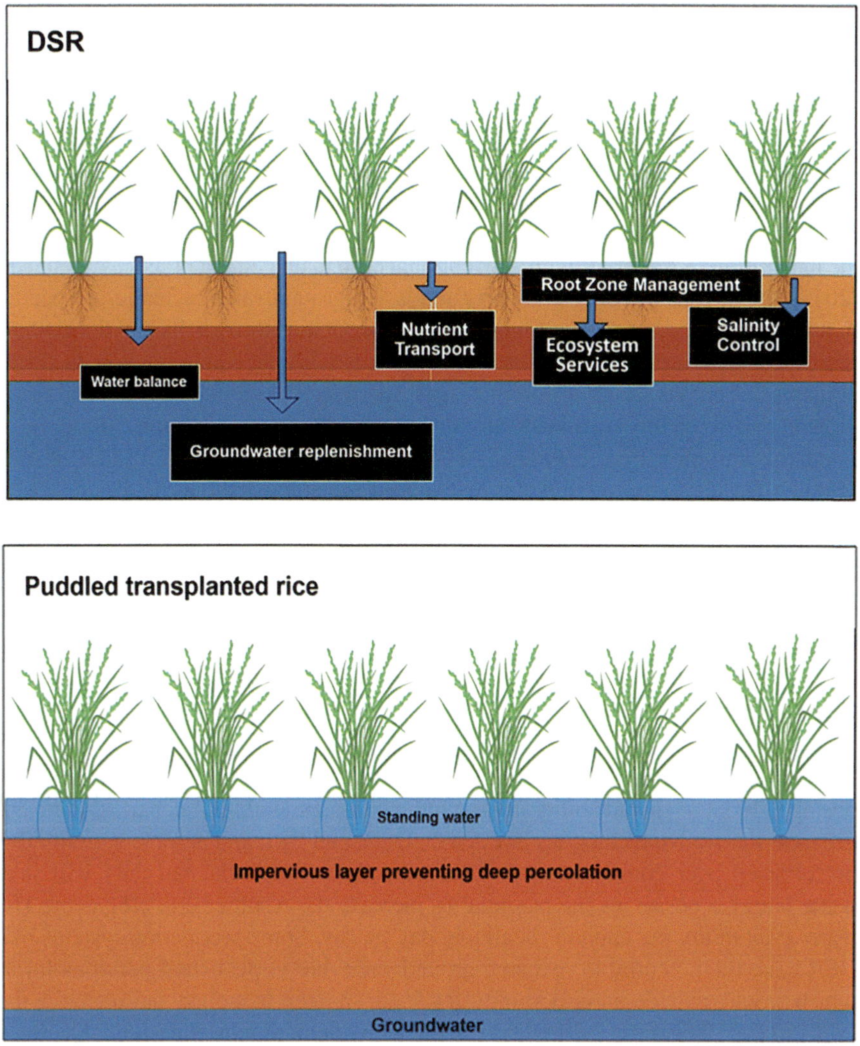

Fig. 9.2 The mechanism of deep percolation in DSR vs. puddled condition. The benefits of percolation in direct-seeded rice in a non-puddled soil (**a**). The deep percolation is prevented in puddled transplanted rice due to formation of impervious layer resulted from puddling (**b**)

During the dry season, reduced rainfall and heightened irrigation water extraction can diminish groundwater levels, resulting in lower recharge rates. Effective water management strategies need to consider these seasonal changes to optimize aquifer recharge and ensure the sustainability of groundwater resources year-round. For instance, implementing rainwater harvesting and managed aquifer recharge techniques during the wet season can capture excess water, while promoting water-saving irrigation practices during the dry season can help mitigate groundwater depletion.

Agricultural Managed Aquifer Recharge (Ag-MAR)

Agricultural managed aquifer recharge (Ag-MAR) is an innovative strategy for applying large volumes of water to agricultural lands, with significant potential for global adoption. Levintal et al. (2023) identified six critical factors for implementing Ag-MAR systems: water sources, soil and unsaturated zone processes, groundwater impacts, crop system compatibility, climate change considerations, and social and economic viability. Crops typically grown in flooded conditions may enhance the integration of Ag-MAR with conventional crop management practices. Kennedy (2015) found that flooding cranberry fields for groundwater recharge at specific times resulted in substantially higher recharge rates than those achieved during harvest flooding. The practice of winter flooding rice fields is gaining traction due to its agronomic benefits, such as promoting straw decomposition, suppressing weed growth, and reducing erosion. Moreover, when well executed, winter flooding of rice fields can offer hydrologic and environmental advantages (Negri et al. 2020). To maintain elevated groundwater levels until the start of the agricultural season (typically from late April to early May in Mediterranean climates), winter flooding of rice fields may need large contiguous areas and extended durations, ideally concluding just before the agricultural season begins (Mayer et al. 2019; Natuhara 2013; Negri et al. 2020).

Conclusion

The climate change and injudicious use of water in water-loving crops like rice are posing a threat to sustainability in near future. To cope with the scenarios, there are different innovative techniques like AWD, DSR, and SRI, and aerobic system must be followed over conventional puddled transplanted rice that not only will save water but also possesses the potential to increase the resource-use efficiency. The water availability per capita is declining day by day. Some production systems like DSR have the potential to recharge groundwater due to persisting pervious layer, which would provide an opportunity for water to deep percolate and recharge the groundwater. Besides, Ag-MAR has large-scale potential to replenish the localized groundwater with its multifaceted approach.

References

Abd El-Latif KM, Abdullahb R (2018) Rice yield and water saving in the system of rice intensification (SRI). Methods 80(120):20–30

Balasubramanian V, Hill JE (2002) Direct seeding of rice in Asia: emerging issues and strategic research needs for the 21st century. In: Pandey S, Mortimer M, Wade L, Tuong TP, Lopez K, Hardy B (eds) Direct seeding: research strategies and opportunities. Los Baños, International Rice Research Institute, pp 15–42

Bethune MG, Selle B, Wang QJ (2008) Understanding and predicting deep percolation under surface irrigation. Water Resour Res 44(12). https://doi.org/10.1029/2007WR006380

Bhatt R, Hossain A, Busari MA, Meena RS (2021) Water footprint in rice-based cropping systems of South Asia. In: Agroecological footprints management for sustainable food system. Springer, Singapore, pp 273–308

Bhushan L, Ladha JK, Gupta RK, Singh S, Tirol-Padre A, Saharawat YS, Gathala M, Pathak H (2007) Saving of water and labor in a rice–wheat system with no-tillage and direct seeding technologies. Agron J 99(5):1288–1296

Biswas A, Mailapalli DR, Raghuwanshi NS, Singh R (2021) Water use efficiencies, productivities, and footprints of rice under a system of rice intensification practice. ACS Agric Sci Technol 1(3):262–269

Boruah P, Sarma A, Das KN (2018) Water saving strategy in rice by alternate wetting and drying technology. Int J Curr Microbiol App Sci 7(3):1333–1340

Bouman BAM (2007) Water management in irrigated rice: coping with water scarcity. International Rice Research Institute, Los Baños

Bouman B (2009) How much water does rice use? Rice Today 8:28–29

Bouman BAM, Lampayan RM, Tuong TP (2007a) Water management in irrigated rice: coping with water scarcity. International Rice Research Institute, Los Baños

Bouman BA, Humphreys E, Tuong TP, Barker R (2007b) Rice and water. Adv Agron 92:187–237

Cai Z, Tsuruta H, Gao M, Xu H, Wei C (2003) Options for mitigating methane emission from a permanently flooded rice field. Glob Chang Biol 9(1):37–45

Chapagain AK, Hoekstra AY (2011) The blue, green and grey water footprint of rice from production and consumption perspectives. Ecol Econ 70(4):749–758

Cheng H, Shu K, Zhu T, Wang L, Liu X, Cai W, Qi Z, Feng S (2022) Effects of alternate wetting and drying irrigation on yield, water and nitrogen use, and greenhouse gas emissions in rice paddy fields. J Clean Prod 349:131487

Choudhary R, Prem G, Kumar A, Singh U, Jat HS, Yadav AK (2016) Comparative study on productivity and profitability of rice (Oryza sativa L.) under different crop establishment methods in Ambala, Haryana. Progress Agric 16(2):183–189

Choudhury BU, Singh AK, Bouman BAM, Prasad J (2007) System of Rice Intensification and irrigated transplanted rice: Effect on crop water productivity. J Indian Soc Soil Sci 55(4):464–470

Dawe D (2005) Increasing water productivity in rice-based systems in Asia: past trends, current problems, and future prospects. Plant Prod Sci 8:221–230

Farooq MKHM, Siddique KH, Rehman H, Aziz T, Lee DJ, Wahid A (2011) Rice direct seeding: experiences, challenges and opportunities. Soil Tillage Res 111(2):87–98

Faruki MRI, Ali MH, Saha RC, Roy AK (2011) Effect of water saving technology through alternate wetting and drying for boro rice cultivation. J Agrofor Environ 5(1):11–14

Gill MS, Kumar P, Kumar A (2006) Growth and yield of direct-seeded rice (Oryza sativa) as influenced by seeding technique and seed rate under irrigated conditions. Indian J Agron 51:283–287

Goitom B, Tripathi RP, Ogbazghi W, Weldeslassie T (2016) Effect of puddling and compaction on water requirements of rice at Hamelmalo, Eritrea. Comput Water Energy Environ Eng. 5(02):27

Gupta RK, Naresh RK, Hobbs PR, Jiaguo Z, Ladha JK (2003) Sustainability of post-green revolution agriculture. The rice–wheat cropping systems of the Indo-Gangetic Plains and China. In: Improving the productivity and sustainability of rice–wheat systems: issues and impacts. ASA Special Publication 65, Madison: WI, pp xx–xx

Hang X, Danso F, Luo J, Liao D, Zhang J, Zhang J (2022) Effects of water-saving irrigation on direct-seeding rice yield and greenhouse gas emissions in north China. Agriculture 12(7):937

Hossain MI, Sufian MA, Hossain ABS, Meisner CA, Lauren JG, Duxbury JM (2003) Performance of bed planting and nitrogen fertilizer under rice–wheat–mungbean cropping systems in Bangladesh. Wheat Research Centre Internal Review Reports. Bangladesh Agricultural Research Institute, Dinajpur, Bangladesh

Humphreys E, Kukal SS, Christen EW, Hira GS, BalwinderSingh S-Y, Sharma RK (2010) Halting the groundwater decline in north-west India—Which crop technologies will be winners? Adv Agron 109:155–217

Ishfaq M, Akbar N, Anjum SA, Anwar-Ijl-Haq M (2020a) Growth, yield and water productivity of dry direct seeded rice and transplanted aromatic rice under different irrigation management regimes. J Integr Agric 19(11):2656–2673

Ishfaq M, Farooq M, Zulfiqar U, Hussain S, Akbar N, Nawaz A, Anjum SA (2020b) Alternate wetting and drying: a water-saving and ecofriendly rice production system. Agric Water Manag 241:106363

Jehangir WA, Masih I, Ahmed S, Gill MA, Ahmad M, Mann RA, Chaudhary MR, Qureshi AS, Turral H (2007) Sustaining crop water productivity in rice-wheat systems of South Asia: a case study from the Punjab, Pakistan, vol 115. Colombo, International Water Management Institute

Joshi E, Kumar D, Lal B, Nepalia V, Gautam P, Vyas AK (2013) Management of direct seeded rice for enhanced resource-use efficiency. Plant Knowl J 2(3):119–134

Kambale J, Singh DK, Sarangi A (2017) Impact of climate change on groundwater recharge in a semi-arid region of northern India. Appl Ecol Environ Res 15(1):335–362

Kamboj BR, Kumar A, Bishnoi DK, Singla K, Kumar V, Jat M L., Chaudhary N, Jat HS, Gosain DK, Khippal A, Garg R, Lathwal OP, Goyal SP, Goyal NK, Yadav AK, Malik DS, Mishra A Bhatia R (2012) Direct Seeded Rice Technology in Western Indo-Gangetic Plains of India: CSISA Experiences. CSISA, IRRI and CIMMYT

Katambara Z, Kahimba FC, Mahoo HF, Mbungu WB, Mhenga F, Reuben P, Maugo M, Nyarubamba A (2013) Adopting the system of rice intensification (SRI) in Tanzania: a review. Agric Sci 2013. https://doi.org/10.4236/as.2013.48053

Kaur J, Singh A (2017) Direct seeded rice: prospects, problems/constraints and researchable issues in India. Curr Agricu Res J 5(1):13

Kennedy CD (2015) Hydrologic and nutrient response of groundwater to flooding of cranberry farms in southeastern Massachusetts, USA. J Hydrol 525:441–449

Kroes J, van Dam J, Supit I, De Abelleyra D, Verón S, de Wit A, Boogaard H, Angelini M, Damiano F, Groenendijk P, Wesseling J, Veldhuizen A (2019) Agrohydrological analysis of groundwater recharge and land use changes in the Pampas of Argentina. Agric Water Manag 213:843–857

Kumar V, Ladha JK (2011) Direct seeding of rice: recent developments and future research needs. Adv Agron 111:297–413

Kumar KA, Rajitha G (2019) Alternate wetting and drying (AWD) irrigation-a smart water saving technology for rice: a review. Int J Curr Microbiol App Sci 8(3):2561–2571

Kumar V, Ladha JK, Gathala MK (2009) Direct drill-seeded rice: a need of the day. In: Annual Meeting of Agronomy Society of America, Pittsburgh, November 1–5, 2009. http://a-cs.confex.com/crops/2009am/webprogram/Paper53386.html

Kumar A, Kumar S, Dahiya K, Kumar S, Kumar M (2015) Productivity and economics of direct seeded rice (*Oryza sativa* L.). J Appl Nat Sci 7(1):410–416

Kumar V, Jinger D, Parthiban P, Lokesh S (2018) Aerobic rice cultivation for enhanced water use efficiency. Indian Farming 68(6):3–6

LaHue GT, Linquist BA (2019) The magnitude and variability of lateral seepage in California rice fields. J Hydrol 574:202–210

LaHue GT, Linquist BA (2021) The contribution of percolation to water balances in water-seeded rice systems. Agric Water Manag 243:106445

Lampayan RM, Rejesus RM, Singleton GR, Bouman BA (2015) Adoption and economics of alternate wetting and drying water management for irrigated lowland rice. Field Crop Res 170:95–108

Lantican MA, Lampayan RM, Bhuiyan SI, Yadav MK (1999) Determinants of improving productivity of dry-seeded rice in rainfed lowlands. Exp Agric 35:127–140

Levintal E, Kniffin ML, Ganot Y, Marwaha N, Murphy NP, Dahlke HE (2023) Agricultural managed aquifer recharge (Ag-MAR)—a method for sustainable groundwater management: a review. Crit Rev Environ Sci Technol 53(3):291–314

Li G, Zhang C, Huo Z (2023) Reconciling crop production and ecological conservation under uncertainty: a fuzzy credibility-based multi-objective simulation-optimization model. Sci Total Environ 873:162340

Liu Y, Liu W, Geng X, Liu B, Fu X, Guo L, Bai J, Zhang Q, Geng Y, Shao X (2022) Direct-seeded rice reduces methane emissions by improving root physiological characteristics through affecting the water status of paddy fields. Rhizosphere 24:100628

Mallareddy M, Thirumalaikumar R, Balasubramanian P, Naseeruddin R, Nithya N, Mariadoss A et al (2023) Maximizing water use efficiency in rice farming: a comprehensive review of innovative irrigation management technologies. Water 15(10):1802

Mann RA, Ashraf M (2004) Improvement of Basmati and its production practices in Pakistan. In: Duffy R et al (eds) Specialty rice of the world: breeding, production and marketing. Food and Agricultural Organisation of the United Nations, Rome. pp 134, 135, 137, 140

Mayer A, Rienzner M, Cesari de Maria S, Romani M, Lasagna A, Facchi A (2019) A comprehensive modelling approach to assess water use efficiencies of different irrigation management options in rice irrigation districts of northern Italy. Water 11(9):1833

Mohammad A, Sudhishri S, Das TK, Singh M, Bhattacharyya R, Das A, Jhanna M, Sharma VK, Dwivedi N, Kumar M (2018) Water balance in direct-seeded rice under conservation agriculture in North-western Indo-Gangetic Plains of India. Irrig Sci 36(6):381–393

Nachshon U (2018) Cropland soil salinization and associated hydrology: trends, processes and examples. Water 10(8):1030

Natuhara Y (2013) Ecosystem services by paddy fields as substitutes of natural wetlands in Japan. Ecol Eng 56:97–106

Nazarieh F, Ansari H, Ziaei AN, Izady A, Davari K, Brunner P (2018) Spatial and temporal dynamics of deep percolation, lag time and recharge in an irrigated semi-arid region. Hydrogeol J 26(7):2507–2520

Negri C, Chiaradia E, Rienzner M, Mayer A, Gandolfi C, Romani M, Facchi A (2020) On the effects of winter flooding on the hydrological balance of rice areas in northern Italy. J Hydrol 590:125401

Neue HU (1993) Methane emission from rice fields. Bioscience 43(7):466–474

Nie L, Peng S, Chen M, Shah F, Huang J, Cui K, Xiang J (2012) Aerobic rice for water-saving agriculture. A review. Agron Sustain Dev 32:411–418

Norton GJ, Shafaei M, Travis AJ, Deacon CM, Danku J, Pond D, Cochrane N, Lockhart K, Salt D, Zhang H, Dodd IC, Hossain M, Islam MR, Price AH (2017) Impact of alternate wetting and drying on rice physiology, grain production, and grain quality. Field Crop Res 205:1–13

Omwenga KG, Mati BM, Home PG (2014) Determination of the effect of the system of rice intensification (SRI) on rice yields and water saving in Mwea Irrigation Scheme, Kenya. J Water Resour Protect 2014. https://doi.org/10.4236/jwarp.2014.610084

Panda S, Majhi PK, Anandan A, Mahender A, Veludandi S, Bastia D, Singh SK, Saha S, Ali J (2021) Proofing direct-seeded rice with better root plasticity and architecture. Int J Mol Sci 22(11):6058

Pandey C, Shukla S, Obreza TA (2007) Development and evaluation of soil moisture-based seepage irrigation management for water use and quality. J Irrig Drain Eng 133(5):435–443

Pathak H, Tewari AN, Sankhyan S, Dubey DS, Mina U, Singh VK, Jain N (2011) Direct-seeded rice: potential, performance and problems-A review. Curr Adv Agric Sci (An International Journal) 3(2):77–88

Patle GT, Singh DK, Sarangi A, Sahoo RN (2017) Modelling of groundwater recharge potential from irrigated paddy field under changing climate. Paddy Water Environ 15(2):413–423

Patra PS, Haque S (2011) Effect of seedling age on tillering pattern and yield of rice (Oryza sativa L.) under system of rice intensification. ARPN J Agric Biol Sci 6(11):33–35

Peng SB, Bouman B, Visperas RM, Castaneda A, Nie LX, Park HK (2006) Comparison between aerobic and flooded rice in the tropics: agronomic performance in an eight-season experiment. Field Crop Res 96:252–259

Prasad R (2011) Aerobic rice systems. Adv Agron 111:207–247

Reddy VR, Reddy PP, Reddy MS, Raju SR (2005) Water use efficiency: a study of system of rice intensification (SRI) adoption in Andhra Pradesh. Indian J Agric Econ 60(3):458–472

Riaz A, Khaliq A, Fiaz S, Noor MA, Nawaz MM, Mahboob W, Ullah S (2018) Weed management in direct seeded rice grown under varying tillage systems and alternate water regimes. Planta Daninha 36:59

Russo T, Alfredo K, Fisher J (2014) Sustainable water management in urban, agricultural, and natural systems. Water 6(12):3934–3956

Santiago-Arenas R, Hadi SN, Fanshuri BA, Ullah H, Datta A (2019) Effect of nitrogen fertiliser and cultivation method on root systems of rice subjected to alternate wetting and drying irrigation. Ann Appl Biol 175(3):388–399

Satyanarayana A, Thiyagarajan TM, Uphoff N (2007) Opportunities for water saving with higher yield from the system of rice intensification. Irrig Sci 25:99–115

Seiler KP, Gat JR (2007) Groundwater recharge from run-off, infiltration and percolation, vol 55. Springer, Cham

Setyanto P, Pramono A, Adriany TA, Susilawati HL, Tokida T, Padre AT, Minamikawa K (2018) Alternate wetting and drying reduces methane emission from a rice paddy in Central Java, Indonesia without yield loss. Soil Sci Plant Nutr 64(1):23–30

Sharma PK, Bhagat RM (1993) Puddling and compaction effects on water permeability of texturally different soils. J Indian Soc Soil Sci 41(1):1–6

Singh KN, Bhattacharyya HC (1989) Direct seeded rice. Published by Mohan Primlani for Oxford and IBH publishing Co. Pvt. Ltd. New Delhi, pp 135–142

Singh AK, Choudhury BU, Bouman BAM (2002) Effects of rice establishment methods on crop performance, water use and mineral nitrogen. In: Bouman BAM, Hengsdijk H, Hardy B, Bihdraban B, Toung TP, Ladha JK (eds) Proceedings of the International Workshop on Water-wise Rice Production. International Rice Research Institute, Los Baños, pp 237–246

Singh VK, Dwivedi BS, Singh SK, Majumdar K, Jat ML, Mishra RP, Rani M (2016) Soil physical properties, yield trends and economics after five years of conservation agriculture based rice-maize system in north-western India. Soil Tillage Res 155:133–148

Smith BD (1998) The emergence of agriculture, 1st edn. W H Freeman & Co., Scientific American Library, New York, NY

Sriphirom P, Chidthaisong A, Towprayoon S (2019) Effect of alternate wetting and drying water management on rice cultivation with low emissions and low water used during wet and dry season. J Clean Prod 223:980–988

Sumana M, Ullah H, Datta A, Shrestha PR, Shrestha S, Kachenchart B (2019) Effects of water and rice straw management practices on water savings and greenhouse gas emissions from a double-rice paddy field in the Central Plain of Thailand. Eur J Agron 107:18–29

Talebnejad R, Sepaskhah AR (2014) Effects of water-saving irrigation and groundwater depth on direct seeding rice growth, yield, and water use in a semi-arid region. Arch Agron Soil Sci 60(1):15–31

Tanaka K, Funakoshi Y, Hokamura T, Yamada F (2010) The role of paddy rice in recharging urban groundwater in the Shira River Basin. Paddy Water Environ 8:217–226

Thakur AK, Mohanty RK, Patil DU, Kumar A (2014) Impact of water management on yield and water productivity with system of rice intensification (SRI) and conventional transplanting system in rice. Paddy Water Environ 12(4):413–424

Tuong TP, Bouman BA (2003) Rice production in water-scarce environments. Water productivity in agriculture: Limits and opportunities for improvement. 1:13–42

Uphoff N (2003) Higher yields with fewer external inputs? The system of rice intensification and potential contributions to agricultural sustainability. Int J Agric Sustain 1(1):38–50

Uphoff N, Randriamiharisoa R (2002) Reducing water use in irrigated rice production with the Madagascar System of Rice Intensification (SRI). Water-wise rice production, pp 71–87

Vijayakumar M, Ramesh S, Chandrasekaran B, Thiyagarajan TM (2006) Effect of system of rice intensification (SRI) practices on yield attributes yield and water productivity of rice (Oryza sativa L.). Res J Agric Biol Sci 2(6):236–242

Vijayakumar S, Nithya N, Saravanane P, Mariadoss A, Subramanian E (2023) Revolutionizing Rice Farming: Maximizing Yield with Minimal Water to Sustain the Hungry Planet

Wang X, Zhang G, Xu YJ (2014) Spatiotemporal groundwater recharge estimation for the largest rice production region in Sanjiang Plain, Northeast China. J Water Supply Res Technol AQUA 63(8):630–641

Waris A, Kumar RM, Surekha K, Meera SN, Nirmala B, Kumar SA (2019) Climate resilient production practices: extent of adoption and barriers faced by rice farmers in Telangana state of India. J Cereal Res 11(3):293–299

Yadav S, Humphreys E, Kukal SS, Gill G, Rangarajan R (2011) Effect of water management on dry seeded and puddled transplanted rice Part 2: water balance and water productivity. Field Crops Res 120(1):123–132

Yao F, Huang J, Cui K, Nie L, Xiang J, Liu X et al (2012) Agronomic performance of high-yielding rice variety grown under alternate wetting and drying irrigation. Field Crop Res 126:16–22

Zhang H, Xue Y, Wang Z, Yang J, Zhang J (2009) An alternate wetting and moderate soil drying regime improves root and shoot growth in rice. Crop Sci 49(6):2246–2260

Zhao L, Wu L, Li Y, Lu X, Zhu D, Uphoff N (2009) Influence of the system of rice intensification on rice yield and nitrogen and water use efficiency with different N application rates. Exp Agric 45(3):275–286

Adoption and Scaling Regenerative Agriculture: Strategies for Sustainable Resource Management in Rice-Based Systems

10

Saikumar C. Bharamappanavara, Yash Srivastava, Biswaranjan Baraj, and Ajay Kumar Mishra

Introduction

Deep-Rooted Practice of High External Input (Chemical-Based) Farming

"The nation that destroys its soil destroys itself," Franklin D. Roosevelt (US President, 1933–1945). Rhodes (2012) captures perfectly the importance of soil and its importance for agriculture and its overall implication for any nation if not managed properly. Every country in its growth cycle starts with developing a strong agricultural base, as it provides livelihood opportunities and fulfills the food needs of the population. The concern over increasing food needs came to the limelight after the 1800s when the global population hit 1 billion for the first time in history and started witnessing an exponential rise, reaching close to 8 billion by 2023 in just two centuries.

This population rise is driven by Asia, which contributes more than half of this roughly 4.5 billion over the same period (Ritchie 2024). This concern got revised attention in the post-World War II period, when newly formed nations started rebuilding, and agriculture was their primary growth engine. There was an evident shift towards chemical-based agriculture, as it promised better yield and production. Hence, the agronomists of the twentieth century were mostly concerned with the need to increase agricultural production and productivity to feed the rising population and provide for industrial feeds, given the constraint of limited land availability.

S. C. Bharamappanavara (✉) · Y. Srivastava · B. Baraj
Sambodhi Research and Communications Pvt. Ltd., Noida, Uttar Pradesh, India
e-mail: saikumar@sambodhi.co.in

A. K. Mishra
International Rice Research Institute South Asia Regional Centre, Varanasi, Uttar Pradesh, India

A. K. Mishra et al. (eds.), *Key Drivers and Indicators of Soil Health Management*, https://doi.org/10.1007/978-981-97-7564-4_10

However, this shift towards chemical-intensive agriculture led to the failure of agriculture's ecological services like carbon sequestration, pollination and filtration, and degradation of natural resources, which further impaired the ability of agroecosystems to sustain productions in the future, while on the other hand creating socio-economic challenges with the increased financial burden on farmers (FAO, NSP—Ecological Intensification n.d.-b). Consequently, a study by the UN FAO (2015) showed that globally, only 60 years of harvest is left (FAO, International Year of Soil Conference 2015), due to the continuous degradation of soil.

In the coming decades, many countries started to report alarming residues of agricultural chemicals in soil, water, air, agricultural products, and even human blood. Research suggests that the massive use of chemical fertilizers worldwide has led to the accumulation of contaminants like heavy metals, e.g., arsenic (As), lead (Pb), and mercury (Hg), in agricultural soils. In the USA, as per the US Geological Survey, of 51 major rivers and aquifers, pesticides were detected in 97% of them (Gilliom 2007). Agricultural chemical inputs denote the different types of chemical applications in agricultural production, such as pesticides (including natural and biological pesticides), chemical fertilizers, veterinary drugs, and feed additives, among others.

Given this, in 1972, the US Government implemented the Federal Environmental Pesticide Control Act (FEPCA), which while acknowledging the negative impact of pesticides on human health and the environment tried to regulate their usage (Federal Environmental Pesticide Control Act of 1972 n.d.). This was a trend which got trickled down to other nations as well. The Indian Government postindependence prioritized agricultural development and food security, through its early Five-Year Plans. Even then, during the mid-1960s, India faced one of the worst food crises coupled with almost all the development metrics in the form of GDP growth, unemployment, income levels, and industrial growth at their lowest levels (Mohapatra et al. 2022). Hence, there was a constant dependence on foreign aid from developed countries, but those had complications in the form of political compulsion. Thus, an essential need was felt to focus on food self-sufficiency since the population and demand for food were only supposed to grow further. However, there were two ways to do it: an increase in productivity/yield or agricultural land extension. Since India's land under agrarian cultivation has already been high, the need was felt to increase the yield, which paved the way for the quick adoption of HYVs in India. Thus, supported by the local needs and adopting the globally dominant paradigm, India under the purview of the Green Revolution in the 1960s promoted the use of chemical fertilizers, pesticides, and high-yielding variety (HYV) seeds, for increasing agriculture production to meet its ever-growing food security needs (Raeboline et al. 2019). In India, a major shift to chemical farming was observed during the green revolution, and farming started exploiting land resources and using heavy external inputs started practicing highly productive commercial cultivation. Further, the subsidized availability of chemical fertilizers facilitated the integration of these inputs into traditional farming systems, effectively raising crop yields and transforming Indian agriculture.

However, as years passed, the excessive use of chemicals resulted in diminishing marginal utility, leading to either stagnant or declining agricultural outputs and growing debt for farmers (Gupta et al. 2020). In India, as per the Fertiliser Association of India (FAI) data, the NPK fertilizer usage has increased from 69,800 tons to 32,535,600 tons from 1950–1951 to 2020–2021, which is a jump of more than 450 times (The Fertiliser Association of India- Statistical Database (2023)) with a higher rise in the past few decades.

Hence, the outcomes of this progress can be understood from three different domains, environmental, economic, and social. In the domain of environment and ecology, synthetic fertilizers have led to soil degradation, diminishing fertility, and increased soil erosion while also causing water pollution and disturbing aquatic ecosystems due to runoffs. Further, there is a rising concern about these chemicals getting into the food chain and having direct consequences on human health. On the economic side, while there was high chemical usage and high input costs for farmers, productivity/yield either stagnated or declined in recent times (Ghosh 2004). Further market uncertainties coupled with the vagaries of nature placed farmers in a dire situation. Hence, the financial implication of this on the farming community created a necessity for institutional intervention to deal with it all at once. Lastly, on the social front, the farming community in general, which is already down the pyramid of development, where women have a major stake in farming with a high involvement of backward castes (Chayal et al. 2013), puts them down further, creating a necessity for intervention in this domain.

Soon the need was felt to not view agriculture merely from a food production perspective but rather more from a multidimensional lens that provides comprehensive services at the societal and national levels. From this, one of the ideas that emerged was ecological intensification, which as per the Food and Agriculture Organization (FAO) is defined as a "knowledge-intensive process that requires optimal management of nature's ecological functions and biodiversity to improve agricultural system performance, efficiency, and farmers' livelihoods" (FAO, Changing paradigms of agriculture n.d.-a). For areas with high to very high agricultural productivity based on high chemical inputs, the idea is to reduce reliance on such inputs while maintaining high levels of productivity and restoring agricultural ecological services. Similarly, for areas with low productivity levels, the idea is to increase it without increasing the overall chemical inputs.

However, presently in India, this whole shift from chemical-based to natural input-based agriculture has been either a farmer or a civil society movement, where these challenges were understood, and they took note of this change with an active acknowledgment of benefits associated with traditional ways of agriculture. However, governments and policymakers have started to take note and become active proponents of this where the Central Indian Government launched a scheme titled Paramparagat Krishi Vikas Yojana in 2015–2016. Further, several state governments took initiatives under the centrally sponsored scheme, for example Andhra Pradesh Community Managed Natural Farming (APCNF) in Andhra Pradesh or other similar interventions in other states across India. Thus, recognizing the urgency of addressing agriculture-related environmental degradation and socio-economic imbalances, global and national policymakers, donor agencies, and stakeholders are increasingly advocating for sustainable agricultural practices (SAPs). Thus, zero-budget natural farming (ZBNF), agroecological approaches, integrated pest management, and organic farming to name a few emerged as viable alternatives to reduce chemical input usage and foster ecological balance.

The sections ahead going forward discuss about few good examples in the form of case studies to present how this whole idea of sustainable agricultural practices (SAPs) is a good example to counter the challenges of agriculture discussed until now.

Rice Cultivation Systems in India

Rice is one of the major crops grown in India, where it is grown in the largest land area, around 44 million hectares, which is 20% of the production area worldwide, and the largest produced crop with a production of 114.5 million tonnes, spread across both the cropping seasons in the country (Agricultural Statistics at a Glance 2022). This production is majorly spread across the Gangetic Plains covering the states of Punjab, Haryana, Uttar Pradesh, Bihar, and West Bengal in the north and other delta and coastal regions in the south for states like Odisha, Andhra Pradesh, and Karnataka. Since rice is an important staple food in India, its production has been enhanced over the years through the use of HYVs, better irrigation facilities, pest and nutrient management practices, and concentrated institutional support. However, this growth in the last decade has almost stagnated due to overexploitation of natural resources, as well as overusage of chemical inputs followed by improper irrigation practices.

Hence, it is important to understand the challenges around the cultivation of rice based on which necessary steps can be taken to address them.

1. ***High usage of water resources***: India is the most populous country in the world, and the requirement for water resource management is already a necessity. However, agricultural systems, like rice, which are heavily dependent on water further add to this pressure. For example, it requires 3000–5000 L of water to produce 1 kg of rice (Kumar et al. 2021). This situation is worsened by the fact that roughly 85–90% of water is solely extracted for agricultural purposes in India. As per the World Bank report, India is already the largest consumer of groundwater in the world (World Bank Group 2012). Hence, given the uncertainty of monsoons lately followed by impacts of climate change in the forms of heat waves and droughts, water availability is supposed to go down further impacting rice cultivation, and thus a concentrated effort is needed to come up with solutions.
2. ***Depletion in natural resources:*** The declining soil health, due to the high-intensity usage of chemicals for the production of rice, has led to increased greenhouse gas emissions from the rice fields. This puts a question on India's

Nationally Determined Contribution (NDC) target to reduce the emission intensity of the GDP to 45% by 2030 from the 2005 level.

3. ***Increased demand for food crops:*** Since India has already surpassed China to become the most populous country, the demand for staple food is only going to increase in the times ahead.

   Hence, this is supposed to further put pressure on the rice cultivation systems to increase already stagnated yields, which, due to lack of knowledge and unawareness among farmers, might lead to further increased usage of chemicals, turning into a counterproductive agricultural system.
4. ***Increased contamination of the surrounding environment:*** In case of an increased application of the chemicals, the runoff is supposed to increase further which has the potential to contaminate the surrounding aquifers as well as lead to groundwater contamination. This further harms the species that are part of the agricultural systems along with the ones that are part of the overall ecosystem, diminishes the ecosystem services that the agricultural system provides, and hampers the regenerative properties. This leads to the biomagnification of chemicals in the food chain as well as positively affects the menace of the weeds on the agricultural field.

Thus, all these challenges need concentrated efforts to be made at the institutional level to look for solutions, which account for both the existing challenges and our preparation for the upcoming ones that might come up. Given this, one popular school of thought is to give up on chemical usage and move back to the more traditional and chemical-free forms of agriculture, which has already been discussed in detail in the previous section.

Hence, through the case studies given ahead, we try to take note of the existing examples that have been a success story in this transition from chemical-based agriculture to a more sustainable form of agriculture and produced positive outcomes at the same time.

Case Study 1: Natural Farming Interventions as the Community-Based Approach in Andhra Pradesh

Andhra Pradesh is an important commercial agricultural state in the Indian perspective, contributing 12% of total rice production, 44% of chili, and 8% of cotton production in the country (Velayutham and Damodaran 2015). This is driven by the state's diverse agroclimatic conditions and fertile land, which is fed by rivers like Godavari and Krishna, thus supporting a variety of crops. This Krishna-Godavari delta region also helps it earn the title of the "Rice Bowl of India" (Niti Aayog: Natural Farming n.d.). Thus, the agricultural sector plays a vital role in Andhra Pradesh's economy, contributing significantly to the state's gross value added (GVA) where, in the year 2021–2022, it alone contributed around 15% to the GVA (Agricultural Statistics at a Glance, 2022–2023, Andhra Pradesh n.d.). However, the sector also engages disproportionately 62% of the working population in agriculture and allied activities.

Among the land utilized in Andhra Pradesh, the most prominent is agricultural land usage where the net sown area is 55.36 lakh hectares, which is roughly 34% of the land in the state (Agricultural Statistics at a Glance, 2022–2023, Andhra Pradesh n.d.). Given that paddy is the most prominent crop, it is important to deep dive since challenges with paddy would speak for the overall challenges with the farming community. It is the most prominent food grain in Andhra accounting for 75% of the overall food grain production in 2022–2023, at 126.20 lakh tonnes. This was practiced on 21.29 lakh hectares during the same period, thus dominating the overall land usage (Season and Crop Coverage Report, Andhra Pradesh 2023). Hence, agriculture is central to the state's politics, economics, and social identity, and paddy is a pivotal component of it. Owing to the centrality of agriculture in the state, the challenges in the sector should be dealt with properly, failing which, the farming community will face enormous difficulties, while the agriculture-based economic development will not be sustained.

The total geographical area of Andhra Pradesh as of 2022–2023 is 162.97 lakh hectares. Despite its prominence in agricultural production and highest rice productivity, the state of Andhra Pradesh is also one of the leading states in the use of chemicals, fertilizers, and pesticides in all commercial crops, and specifically in the cultivation of rice as well, and has observed the following specific challenges.

Challenges in Agriculture in Andhra Pradesh

1. Andhra Pradesh agriculture is largely reliant on rainfall, with the gross irrigated area at 54.2% as of 2022–2023 (Agricultural Statistics at a Glance, 2022–2023, Andhra Pradesh n.d.), thus leaving roughly 50% population prone to the vagaries of monsoon and other seasonal conditions. Farmers repeatedly face prolonged dry spells and drying of borewells, coupled with severe droughts, especially in southern Andhra Pradesh.
2. This problem is exacerbated by the current agricultural practices, which result in continuous loss of soil organic matter, leading to greater soil and water loss. It has been estimated that worldwide, 40–60% of soil organic matter is already lost, and if this continues, the topsoil will be eroded in another 60 years (Xiao 2018). Concomitantly, the water-holding capacity of the soil has been severely diminished.
3. The average landholding size in the state is around 1.2 ha, and 62% of the total land is owned by marginal farmers having less than 1 ha of land (Agricultural Statistics at a Glance, 2022–2023, Andhra Pradesh n.d.). The small holding has put rural HHs in a situation, where the income is neither sufficient to meet the family requirement nor do they have additional employment opportunities at the village level. In addition, there are problems associated with the livelihoods of landless agriculture workers and tenants.
4. The extant agriculture paradigm induces farmers to rely heavily on synthetic fertilizers and pesticides, which has, in turn, led to several problems ranging from exceedingly high costs of cultivation to an alarming rate of soil erosion on

one side, and saturated productivity on the other side. Given this, like the global agricultural paradigm of chemical-based farming, the same ails the state's farming, with consumption of fertilizer being as high as 225 kg/ha, which is the second highest in the country only after Punjab.

5. Since one-third of the land falls under the arid dry zone in Andhra Pradesh, which is a rain shadow area, it creates challenges in terms of the availability of water for irrigation since rainfall is scanty and unpredictable (Andhra Pradesh - General Scenario and Current Trend n.d.). All these problems feed into challenges faced by the farming community, where, as per the NSSO data, Andhra Pradesh has the highest number of indebted agricultural households in India, i.e., 92.9% of agricultural households. Further, as per the NCRB's Accidents and Suicides in India, 2021 report, a total of 1065 agricultural workers (farmers, tenant farmers, and others) committed suicide in 2021, which is third highest in the country, just behind Maharashtra and Karnataka (Andhra Pradesh: 1065 farmers ended lives in 2021, third highest, 2022).

As is evident, the nature of the problem is highly complex, with numerous interrelated parts feeding into each other. The foregoing aspects, along with other contributory reasons, have resulted in ecological deterioration, and farmers are being perpetually trapped in debt. These trends highlight the challenges faced by the agricultural sector in Andhra Pradesh, emphasizing the need for effective water management strategies and sustainable agricultural practices to ensure the resilience and productivity of this essential economic sector. As in the rest of the country, the agriculture sector in AP is plagued by a host of systemic and nonsystemic problems. At the beginning of the decade, the concept of sustainable agricultural practices was started slowly by a few village organizations and self-help group (SHG) members and had witnessed a clear impact in terms of reduced cost of cultivation and increased yield even with the minimal intervention and shift to natural farming principles. The success story has attracted stakeholders in the state and has been scaled up gradually over the last 10 years. For 5 years, the focus on scaling up and making better outreach was very high, and sincere efforts have been put through a separate implementation body Rythu Sadhikara Samstha (RySS). The concept has also been received by farmers and has been initially adopted on a trial basis. Initially, farmers start with a combination of conventional practices, and later a complete shift is observed among a few farmers as seed-to-seed adopters completely practice natural farming on their plots. At present, nearly seven lakh farmers are practicing natural farming at different levels on different crops and in different agroclimatic conditions. During these years, RySS has also evolved and dynamically has adopted a few changes to the original natural farming principles, and along with the basic principles of Jeevamrutham, Bheejamrutham, and Kashayas has also focused on and started a not-for-profit company established by the Department of Agriculture, Government of Andhra Pradesh, adopting pre-monsoon dry sowing (PMDS), use of *Azolla* in paddy fields, live mulching, IPM, INM, and minimum tillage practices as part of community-based natural farming. However, all these practices together aim to have sustainable agricultural production in harmony with nature using more in-house inputs, reducing exploitation of natural resources, and enhancing the existing ecosystem services. Presently, with all keen effort and focus put by the Government of Andhra Pradesh, the program has reached out well and has been executed and monitored by a well-organized autonomous institute called Rythu Sadhikara Samstha (RySS), having internal community resource persons (iCRPs) at the village level, Mandal anchors at Mandal level, and district project managers at district level, who are all part of the agriculture department. However, before deep diving into this, it is important to understand the agriculture scenario in the state.

Concept of Community-Based Natural Farming and Its Scaling Scenario

Given the abovementioned challenges, the Government of Andhra Pradesh introduced community-managed natural farming (CNF) in 2015–2016, based on the concept of the zero-budget natural farming model (ZBNF). This program builds onto the state's experiences of non-pesticide management and Community Managed Sustainable Agriculture (CMSA), which was implemented from 2005–2006 to 2012–2013 (Gulati et al. 2024). ZBNF is seen as an alternative to the existing high-cost chemical input-based agriculture, where farmers would shift to sustainable farming practices and use low-cost, locally sourced inputs while eliminating chemical usage to make farming more climate resilient. Thus, in principle, ZBNF, popularly called CNF, aims to reduce the use of external inputs and increase the use of eco-friendly inputs that are locally prepared, thus reducing the cost of cultivation and adding to the profit margins while also bringing the ecological balance by adopting sustainable practices such as crop rotation and PMDS live crop mulching. Thus, CNF helps Andhra Pradesh and the overall Indian agricultural scenario make significant progress towards a quarter of the 169 SDGs tackling hunger, poverty, and inequality; expanding nutritional choices; responding to climate change; and safeguarding biodiversity. Eventually, the Government of Andhra Pradesh plans to scale CNF practices across the entire 8 million hectares in the state by 2027. By principle, the program's approach aims to address both environmental and socio-economic concerns ahead of us in the agricultural sector: (1) reducing the use of external inputs and increasing the use of eco-friendly inputs prepared using local resources, (2) reducing the cost of cultivation and adding to profit margin, and (3) bringing ecological balance by adopting sustainable practices such as crop rotation, pre-monsoon dry sowing (PMDS) 4, and live crop mulching. Concerning scale, experience, and local context, the project has evolved in its principles dynamically and improvised the practices against the base ZBNF principles defined in Table 10.1. Nonetheless, the basic tenet of intervention remains the same—that the farmer should use low-cost, locally sourced inputs and eliminate the use of chemical fertilizers and pesticides to make agriculture more climate resilient. Primarily, there are four practices (or Amruthams) that form the core practice of natural farming: (1) Bheejamrutham (seeds are covered in preparation of cow dung andamp, urine, and other bio-inputs); (2) Jeevamrutham (a concoction of bio-inputs made using cow

Table 10.1 Basic principles of ZBNF (APCNF n.d.)

NF wheel	Amrutham	Description of wheel	Ingredients used	Broader benefits
Seed treatment	Bheejamrutham	Seeds/seedlings are applied/quoted/ dipped in a locally prepared solution before sowing/ transplanting	Cow dung, cow urine, water, bio-inputs, asafoetida, *Phyllanthus emblica*	Effectively controls many of the seed-borne diseases, brings immunity against many of the soilborne infestations Increases germination
Microbial inoculants	Ghana-Jeevamrutham (solid)	Is a soil amendment, used as a bio-inoculum. Incorporated in the topsoil. Adds nutrients to the soil	Cow dung, cow urine, jaggery (unrefined cane sugar), topsoil from the field and any gram (legume) floor	Improves soil microbial activity Increases soil fertility
	Drava-Jeevamrutham (liquid)	Is a liquid amendment and acts as an inoculum. Can be applied through drenching or as a foliar spray	Jaggery (unrefined cane sugar), topsoil from the field, and any gram (legume) floor	Improves soil microbial activity Increases soil fertility Provides direct nutrients to soil
Tonics	Kashaya (Dashaparni, Panchagavya, Agni Astra, Brahmastra)	Foliar spray replacing pesticides, insecticides, and micronutrients	Various locally available ingredients are used in different proportions depending on the tonic	Reduces cost of cultivation by replacing chemical pesticides and insecticides
Mulching	Acchadana: (i) using dry crop residues or (ii) crop/live mulching	Dry crop mulching covers the root zone and conserves moisture Live crop mulching increases crop cover	Growing pulses, vegetable, and fodder crops	Live mulching increases crop diversity and soil health and adds to farm income
Mulching (live/dry)	Acchadana: (i) using dry crop residues or (ii) crop/live mulching	Dry crop mulching covers the root zone and conserves moisture Live crop mulching increases crop cover	Residual management, efficient use of fodder	Improves soil fertility, recycling of resources, and water-holding capacity

dung andamp, urine, jaggery, legume flour, etc. available in solid and liquid form and which can be injected into or sprayed over the soil); (3) Acchadana (mulching using crop residue or live crops to maintain green cover; pre-monsoon dry sowing, a breakthrough practice wherein crops are sown before the monsoon season and in dry periods throughout the year to maintain green cover which acts as live mulching, is vigorously promoted under this practice) and Kashayas (tonics made using cow dung andamp, urine, chili paste/powder, neem leaves, etc. that function as pesticides/insecticides; for more, please see Table 10.1); and (4) Waaphasa (soil aeration), the fourth wheel of NF practice, which is essentially an outcome of the first three wheels observed in the soil in terms of better water-holding capacity, good soil aeration, and increased microbial activities.

One of the major practices taken up apart from the abovementioned one is pre-monsoon dry sowing (PMDS), which helps in the full use of cropland water by plant roots. It increases the water storage in the root zone, reduces the inter-row evaporation, and controls excessive transpiration. This creates a special microclimate advantageous to plant growth, having a synergic relation among the crops grown and development leading to an increase in the crop yield per unit area. This technology has reduced the likelihood of total crop failure due to adverse weather and has given diversification benefits (biological and financial) to the farmers in project areas. The awareness about the above-explained problems existed for previous decades, and hence multiple large-scale initiatives in the state of Andhra Pradesh have been taken in the past to combat them. Two such initiatives are the Society for Elimination of Rural Poverty (SERP) and Community Managed Sustainable Agriculture (CMSA). SERP has facilitated grassroots mobilization in the form of SHGs and village organizations (VOs) at various levels which form the institutional skeleton, while the Centre for Sustainable Agriculture (CSA) has focused directly on furthering natural farming at the community level helping in a smooth transition into CNF. Andhra Pradesh has an extensive network of local institutions, where 152,935 SHGs and 6593 VOs are actively involved in program implementation. Further, there is a focus on women as a point of intervention, with an expectation of a multiplier effect. For example, women form 69% of the internal community resource persons (iCRPs) who are being trained as part of the capacity-building program under the APCNF intervention.

Impact of the CNF Program on Rice Cultivation

A set of panel farmers were selected and tracked over 5 years. The observations were made about the farmers divided between two categories, CNF and non-CNF farmers, while the other was not part of the intervention. Based on the midline study, the following were the findings. The paddy crop, which is the most cultivated crop in both main seasons of Kharif (monsoon season—June to October) and Rabi (winter season—October to February), was further compared between CNF and non-CNF farmers for input utilization, cost of cultivation, output, and yield. Paddy was the main crop of farmers for the Kharif season; the other most grown crops were groundnut and black gram, while for the Rabi season, those were chickpea, maize, and black gram. The cropping intensity in Andhra Pradesh is already very good and almost achieving saturation, as electricity for groundwater is free of cost, and there is considerable use of tube wells by farmers.

Ingredients Used and Cost of Cultivation

Ingredient (raw material) collection and input preparation: As the name indicates, activities under this domain relate to the procurement of input materials and preparation of inputs for largely three purposes: (1) nutrient management, (2) seed treatment, and (3) pest management.

Under conventional agriculture, fertilizers and chemical pesticides are the key inputs and are readily available in input shops, and extension workers largely actively market the same. Natural farmers, however, incur additional labor in not only the collection of ingredients such as neem leaves, cow urine and dung, tobacco powder, chemical-free mud, etc. but also the preparation of final inputs. This is more prevalent in areas where NPM or RBK shops do not market ready-to-use NF inputs. The analysis of cost aids in understanding the major contributors towards reducing the cost of cultivation and in identifying opportunities to bring out efficiency to achieve fair profitability. The following lists out the major contributors to this segment of input costs.

1. **Inter-cultural activities:** Between sowing and final crop harvest, a host of activities (e.g., weeding, intercultural plowing, fertilizer application) are conducted by farmers to ensure optimal output. NF farmers face additional labor on two fronts: firstly, for the same set of activities undertaken by conventional farmers, such as fertilizer application, weeding, and pest management, and secondly, for additional natural farming activities such as mulching, intercropping, and integrated pest management (pheromone traps, mechanical methods, trap crops, etc.).
2. **Minimum tillage operations:** It is also useful to contextualize the role of tillage practice and machinery. An important principle under natural farming is that the soil must be disturbed as minimally as possible. This directly translates to reduced (conservation) tillage or zero tillage and is one of the key areas in which there is a difference in machinery used by CNF and non-CNF farmers, leading to a reduction in the cost of cultivation in CNF practice. Further, for farmers who practice intercropping, using machinery for crop harvest is difficult for manual harvest.
3. **Integrated pest management (IPM):** IPM refers to a sustainable agricultural practice of managing pests by combining biological, cultural, physical, and chemical tools in a way that minimizes economic, health, and environmental risks. The stem borers that afflict rice are by far the biggest pest, while other pest attacks included are the plant hoppers, spotted worms, armyworms, aphids, thrips, and leafhoppers. To counter these pests, the usage of solutions like Bheejamrutham, Kashayas, and neem extracts is seen as an effective way fol-

lowed by the usage of pheromone traps and trap crops. Apart from pest management, these solutions were found to be effective against diseases like wilt, leaf spot, rust, and mosaic. Thus, active adoption of these IPM methods by farmers provides an overall improvement in agricultural practices.

4. **Improved nutrient management (INM):** In addition to pest management, nutrient management is essential for improved crop productivity and reducing diseases in the crops. To promote sustainable nutrient management, soil testing comes as an essential step for understanding the nutrient composition and nutrient management of the soil. Among the methods promoted, one of the common ones which was also traditionally followed was intercropping, where complementary crops are sown alongside. Adoption of all such measures and balanced application of fertilizers lead to specifically greater and efficient uptake of nutrients, and further intercropping and usage of farmyard manure and legumes have led to minimal deterioration of soil; reduction in nutrient losses to ground and surface water bodies and atmosphere; improved physical, chemical, and biological functioning of soil; and provision of balanced nutrition to crops leading to enhanced crop yields. Further, practicing INM along with bio-fertilizers and inoculants with augmentation of pulse areas for nitrogen fixation (better nutrient recycling arresting nutrient loss, reduced emission) is an excellent example of adaptation with mitigation co-benefits.
5. **Reduced cost of cultivation and increased income:** The financial analysis of CNF and SAPs highlights the economic returns for rice cultivation during Kharif and Rabi seasons, evidencing reduced cost of cultivation as well as enhanced income, compared to conventional cultivation. Upon examining the application of nutrient management (fertilizers and manures) (Praveen 2019), it must be appreciated that most of the CNF farmers adopt a balanced application of organic manures which includes judicious management of nutrients with the use of Jeevamrutham and Bheejamrutham. While most of the conventional farmers provide due importance to fertilizers, they do so with minimal use of organic fertilizers alongside irrational application of chemical fertilizers. Furthermore, among conventional farmers, nitrogenous fertilizers are the most prominently adopted, with other essential elements restricted to a reduced role. This may lead to reduced costs on manures and fertilizers in the immediate future, but it masks the deteriorating impact on soil fertility, sustainability, and crop productivity in the long run. CNF farmers with SAPs have weaned themselves away from traditional habits of high use of chemicals and fertilizers, shifted to the use of Jeevamruthams and Kashayas, and adopted improved integrated nutrient management (INM) and integrated pest management (IPM) practices.

Case Study 2: Accelerating Farm Incomes Through Adoption of Sustainable Practices

Agricultural practices (SAPs) and market linkages in Telangana, India, as already discussed, have achieved significant improvements in agricultural productivity since the green revolution (John and Babu 2021), helping it achieve sufficiency in food grain production and even contributing to foreign food supplies. Since the last two decades, due to the changing rural scenario, urbanization, and lack of farm labor, the agricultural landscape across India has changed. This has led farmers to shift towards less labor-intensive crops, move towards mechanization, or take up commercial crop cultivation in smaller areas of their total landholding to meet their livelihood needs and by leasing out the remaining piece of their land. These have put farmers under pressure by increased costs of cultivation and reduced profits, despite increased productivity. Since 2016–2017, the Government of India has taken up initiatives to address the changing scenario aimed at Doubling Farmer's Income (DFI). The policy aims to achieve this through (1) rational use of resources, (2) promoting sustainable agricultural practices (SAPs), and (3) demand-oriented crop planning and market linkage. These objectives of SAP have been implemented under the umbrella of different programs/schemes across India by state and central government departments, research institutes, and research for development institutes.

One such intervention in recent times has been the introduction of new pathways to attain the objectives of increasing the productivity of farmers, by focusing on bringing sustainable behavioral changes among farmers through the creation of awareness, training, and capacity building, addressing both sides of the agricultural sector such as sustainable production and viable and efficient market linkages. A few of the best practices adopted in recent times to disseminate technologies to farmers are (i) farmer's field schools and (ii) farm field demonstrations. The most popular SAPs disseminated are urea deep placement (UDP), fertilizer deep placement (FDP), FIBSOL, integrated pest management (IPM), safe vegetable production, improved variety/climate-resilient/certified seeds, mechanization, and market linkages. However, the main challenges observed among the farming community to adopt the SAPs are lack of finance for procuring inputs and lack of differentiated demand for better quality produce. Apart from demonstration and dissemination activities, knowledge sharing and learning are one of the vital approaches for effective outreach of SAPs among farmers, for which farmer scientists and agri-champions are identified among farmers themselves who have better adoption of SAPs and can be used as medium for other farmers to showcase the success stories within their communities. Women and young farmers are to be given special focus for better outreach and wide adoption. The major factors driving the sustainability of SAP practices are availability of required financial support and establishing strong market linkages with private players to help farmers sell better quality produce grown under SAPs at a rate commensurate with the quality, which is the need of the hour. Hence, it is important to integrate success stories, such as agri-champions (AEs) and agri-farmer scientists, with the market by creating forward and backward linkages with key market players, farmer producer organizations (FPOs), and government programs that support entrepreneurship. Market linkages are very important interventions providing a more focused approach in terms of establishing both forward and backward linkages—this will lead to an increase in farm income, because of a decrease in the cost of cultivation, efficient use of resources, and

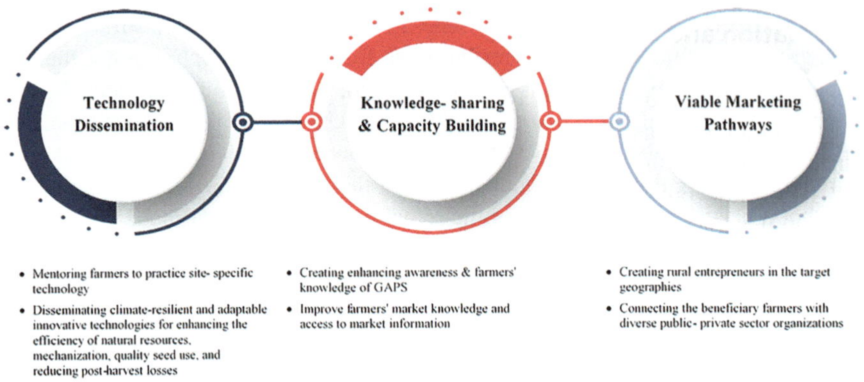

Fig. 10.1 Holistic approach to achieve sustainable rice production and improve productivity concurrently

increased yield which all will lead to increased farm income. Apart from production and market-oriented focus, there is a need to change the mindset of rice-growing farmers about farming and to look at it as agri-entrepreneurship (AE) activities, which has a lot of potential to create a positive business environment in rural areas rather than just a livelihood sustenance activity (Fig. 10.1).

However, these case studies of CNF from AP and SAPs from Telangana become important to understand the relevance of sustainable agricultural practices and natural farming in paddy-based agricultural systems. The major crops under cultivation in AP and Telangana are rice, chili, cotton, wheat, chickpea, sorghum, and millet, and the positive impact it produced with other crops builds an even stronger case for the idea of sustainable agricultural practices and its benefits going forward.

Challenges in Practicing NF

Scholars (Duddigan et al. 2022) have noted that "the APCNF story seems to have worked with the principle: if the small-marginal farmers whose livelihood has taken a beating under 'chemical agriculture' are made to regain their confidence that their knowledge and locally accessible resources could bring stability to their farming with improvements in how and what they produce and consume, then the success of natural farming is ensured." To ensure this success, we find the following to be noteworthy.

1. **Continuity of chemical usage in the field:** Even with the continuous efforts of different state governments to shift the focus of conventional farmers to chemical-free farming practices, there is still a high prevalence of chemical usage. There could be multiple reasons for this, such as habituation with past practices of chemical usage, easy availability of chemical inputs, and lobby of retailers who dominate the agriculture practices and decisions at the village level. Given this,

revised efforts need to be made to make farmers aware of the SAPs, along with increasing their capacity in terms of training for preparing and applying NF inputs, so that adoption can be increased and can eventually replace conventional chemicals.

2. **Not getting a premium price for the NF produce:** One of the motivations that farmers have is to get a premium or a higher price, as is also the popular narrative when promoting natural/organic farming outputs. Not having a guarantee for higher prices also acts as a psychological barrier to the farmers when it comes to investing in something new, given that they are not getting any significant remuneration for the same. Therefore, there is a particular need to promote the certification avenues and also build market linkage channels so that farmers can realize better market value for their produce.
3. **Unavailability of NF input:** An important raw material for the preparation of NF inputs is cow dung and cow urine, apart from many others. However, a lot of the farmers do not have big ruminants, i.e., cow or buffalo or bull or ox. This means that they have to rely on outside sources to get these raw materials, which are essential for ZBNF inputs. This directly impacts the adoption of these practices. The same is true for other inputs like jaggery, which again needs to be readily available. This also adds to the cost of inputs for an agricultural cycle and hence might be a demotivating factor for farmers to actively adopt this shift.
4. **Hindrance in adoption:** When in general, the adoption of SAPs is discussed with farmers, one of the major hindrances discussed is the labor-intensive nature. For example, the preparation of NF inputs or manual weeding requires a lot of labor, which might add to the labor cost and bring the input cost to higher levels while also creating a constant demand that might be difficult to meet at times. Hence, the cost and time implications of the exercise are inherent roadblocks to the adoption.

Scaling-Up and Scaling-Out Strategies for Promoting Sustainable Agricultural Practices in Rice Cultivation

There is an immense need to establish an adept institutional mechanism with vertical and horizontal linkages at various levels, building the capacity of farmers and grassroots-level organizations to adopt sustainable agricultural practices to achieve stability and enhanced productivity and stabilized income in rice cultivation. These linkages on partnerships at different levels of functionaries are recommended to remain accessible and available to beneficiary farmers in the days ahead, ensuring the sustainability of the project's wide-ranging interventions. Despite the impactful results of natural farming practices and sustainable agricultural practices, it is important to come up with scaling-up and scaling-out strategies to ensure long-term sustainability achieving the global SDGs.

1. **Dissemination of SAPs through local institutions through people's participation in planning and implementation:** The local community institutions played a key role within the community in planning, dissemination, outreach, and adoption of SAPs among rice growers. The village-level institutions and various stakeholders, if they spearhead the dissemination of practices and procurement of NF inputs and lead training, workshops, and exposure to execute these intricate yet expansive processes, will augur well for the future concerning the promotion and pursuit of sustainable agricultural practices even after the project's impending exit. In the days ahead, it is envisioned that the local institutions continue to liaison closely with the community institutions, e.g., the village organizations, watershed development committees, farmer interest groups, as well as producer organizations existing in the village, which will keep SAPs and sustainable production as the main mandate of their organization. This will lead to the implementation of activities that are demand driven and technically sound and ensure sustainable agricultural practices in production safeguarding the prosperity of the community.
2. **Inclusion of SAPs as the mandate of the line department and strengthening department staff:** To make SAPs and natural farming practices long sustainable and available to farmers at their doorstep, several functionaries from multiple line departments (agriculture/horticulture/watershed/water resources) have to deeply be involved in planning and implementation. Predominantly, professionals across different teams possessing high technical acumen are expected to continue providing their services and disseminating SAPs to the community at large, thus securing sustainable results and benefits. Further, the concurrent support of many national- and state-level research institutes, extension officials, and additional knowledge gained by the line departments can strengthen institutional memory and will remain an integral part of the action plans of relevant line departments ensuring long-term sustainability.

Suggestions and Way Forward

Given the strong interlinkage between SAPs and sustainable development goals, assessing SAPs, contribution towards the achievement of SDGs becomes critical. Numerous projects, funding agencies, implementation stakeholders, and partners are focusing more and more towards sustainable production practices such as natural farming and on efficient and stable market value chain systems. Many of the programs implemented have proven the scope and potential to increase farmers' income by giving due care for safe and sustainable production using natural farming and SAPs and agri-entrepreneurial activities. There is also a special focus on small and marginal farmers who make up most of the farming community in India, addressing their own set of challenges and issues in farm production activities. The recent experiences have touched upon multiple aspects and have brought many insightful learning. Based on this, the following are a few recommendations/suggestions to take up in the future for scaling up and expanding sustainable production and marketing practices making them accessible and available to marginal and smallholding farmers. Based on these deep insights on the NF and CNF program adoption, practices, and lesson learning, one of the key impacts is reduced cost of cultivation.

Reduced cost of cultivation: With the usage of NF inputs as well as locally prepared Amruthams, Kashayas, compost, and other ways of SAPs, the overall cost of cultivation can be brought down. This is further compounded by the more efficient means of production practices reducing the cost associated with more traditional ways. When the input cost goes down, it has a direct effect on the debt structure of the rural household since most of this debt is taken to account for input cost during the agriculture cycle. This improves the resilience of agricultural house holds as well as enables them to stay afloat financially in case of any economic challenges.

NF and SAPs show clear signs of improvement for the farmers over a period where not only farming households have become more resilient economically but also there is a clear improvement in the surrounding ecology, like aspects of soil health.

1. Ensuring Effective Implementation and Outcomes
   (a) **Bridging the SAP awareness to adoption gaps and increased support**: There is a continuous need for the institutions to plan to take SAP knowledge and reach out to farmers, particularly those who are decision-makers by engaging larger communities. Further, other community platforms such as village organizations, Gram Sabha, and other agricultural institutions such as KVKs and FPO can be leveraged to promote SAPs. Further, expansion and improvement should focus not only on expansion by area and the number of outreaches but also on achieving outcomes and impact through (1) ensuring effective implementation and increased income, (2) empowering field staff and agri-champions through incentives, (3) raising farmers' awareness, (4) leveraging existing SAPs and agri-champions interventions, and (5) facilitating financial instruments.
   (b) **Responsible production and consumption practices:** The SDG-12 (Ensure Sustainable Consumption and Production patterns) target is also to achieve quality production of rice and provide consumers with a quality product of improved productivity as well as increased farmers' income through enhanced resource efficiency and farmer-market linkages. The aim of SDG-12 can be achieved through knowledge sharing and technology dissemination to affect a shift to SAPs promoting integrated soil-seed-water management and responsible production. Further, it aims to create demand for the crops grown by these methods through FFDs and create and enhance market linkages and agri-entrepreneurs, hence promoting sustainable consumption. Any targeted program should focus on three diversified components, i.e., technology dissemination, knowledge sharing and capacity building, and microenterprise development.

(c) **Improve focus on farmer-producer organizations:** Most of the rice-growing farmers are small and marginal farmers, and there is an urgent need for capacity augmentation of these smallholder farmers. One way could be to integrate all these small farmers through FPOs to get the benefit of input supply, production services, and commodity value chains and strengthen the supply chain for seed inputs and then climate-resilient crops produced. It is recommended that there is a great need to support the collectivization of farmers and FPOs to take up sustainable production and pay stringent attention to environmental and social guidelines, participation of women farmers, and expansions in their market-oriented agribusiness activities. It is also recommended that FPOs and extension activities focus on the use of resilient inputs and emphasize more on market-oriented rice production, strengthening rice value chains, establishing sustainable and efficient market linkages, and making quality rice produce available for consumers at a fair price.

(d) **Strengthening partnerships, collaborations, and enhancing stakeholder engagement**: The conceptual plan for the spread and dissemination of natural farming and SAPs should be holistic, from disseminating knowledge to creating awareness, procurement and supply of required inputs, and establishing market linkages. Furthermore, direct involvement or collaborations with technology providers, practitioners, resource persons, and financial agencies would enable more integrated and innovative strategies and create a strong ecosystem for rice and other crops. Thus, there is a clear recommendation for strengthening of partnerships with various stakeholders, including government agencies, nongovernmental organizations (NGOs), research institutions, and the private sector including agri-startups, to facilitate the transfer of knowledge and innovations to farmers. Engaging with NGOs and community-based organizations can enhance the reach and impact of NF and SAP interventions at the grassroots level. Private-sector partners can provide access to advanced technologies, financial resources, and market linkages.

(e) **Establishing market linkages:** The adoption and practice of natural farming and SAPs can be improved and made a sustained practice among rice-growing farmers, only if the produce fetches comparatively better prices in the market. Proper strategies for implementation have to be made, and significant effort has to be put in to establish market linkages for farmers through FPOs and village organizations, to accomplish sustainable linkages. Hence, furthermore, the scope to focus on strengthening and establishing rice (paddy) value chains where farmers can aggregate and sell their produce directly to consumers at better prices is crucial. The effort of FPOs and farmer scientists/agri-champions to establish market linkages will only be effective in having a visible outcome in terms of an increase in income if the linkages are stable in the procurement of SAP-produced rice. Hence, developing service linkages that facilitate direct connections between natural farming practitioners, SAP farmers, and markets will optimize profits and encourage rice-growing farmers to continue using sustainable practices.

This can be done by creating both forward and backward linkages with big market players such as JioMart, BigBasket, and Reliance Fresh as well as with FPOs and government programs that support entrepreneurship.

(f) **Empowering field staff and incentives:** For any scheme or program to be successfully executed, a strong presence of field cadre is pivotal. This has been proved with several government-run schemes like iCRPs in the case of RySS and ZBNF in Andhra Pradesh. It is very much essential to incentivize field-level staff for their effort in promoting and encouraging farmers to undertake sustainable practices in their farms. Performance-based incentives are proposed to be provided to the field staff working with farmers, with performance based bonuses and other incentives such as honorarium and benefits tied to the extent of farmers in their village practicing natural farming and sustainable agricultural practices in their paddy fields, as they are the key drivers of the program. This encourages conversion from demonstration to actual practice. Further, to achieve technical proficiency, it is important to ensure that the program/grassroots staff are well trained in the new advances of natural farming practices and SAPs and in establishing agri-entrepreneurship businesses. This will help them to convert FFS/FFD participation to SAP practice and AE ventures. This can aid in higher achievement of the program objectives.

(g) **Widening existing sustainable practices and natural farming intervention:** Though the present school of thought on natural farming and SAPs is already established with evidence, and accepted and adopted by farming communities, there is a scope to craft and come up with need-based SAPs specifically for rice crops. Hence, the natural farming practices and SAPs must be dynamic and context specific and have to be revisited regularly before dissemination, and a need-based assessment exercise is suggested to be conducted, which will help in crafting a need-based list of NF and SAPs based on the priorities of farmers and to address current challenges. This approach of need-based assessment can convert into a high rate of practice. Context-specific SAP training is vital, as the same set of SAPs may not be the requirement of all the farmers, since they are used based on season, type of variety, kind of production practices adopted/practiced, irrigation sources, and so on. Hence, there is a need for such customization in the implementation of activities. However, the inclusion of more sustainable agricultural practices is a common suggestion across all natural farming and sustainable agricultural practices to cover more topics during training and dissemination through FFS and FFDs and to provide handholding support. Therefore, the inclusion of more need-based topics is recommended as need-based knowledge dissemination will empower rice-growing farmers and encourage long-term engagement and wider adoption of SAPs in rice cultivation.

(h) **Convergence with other government schemes:** To bring effective and efficient dissemination and reach out to smallholding farmers on sustainable agricultural practices, it is vital to explore the convergence with other government schemes which can lead to a multiplier effect towards the target.

Through convergence, they can better achieve financial efficiency, integrated approach, and enhanced impact. By aligning efforts and pooling resources, the project along with the diverse initiatives can work together towards a common goal of enhancing resilience and sustainability in agriculture.

(i) **Digital support for promoting CNF and SAPs:** Since ICT has been an important instrument in disseminating recent innovations to farmers and rural communities, it is important to emphasize a robust information and communications technology (ICT) system to support and disseminate the required support for practicing CNF and SAPs. ICT can help in timely and larger outreach with reasonable financial resources. A robust road map plan for ICT support can bring in a remarkable revolution by providing required training to implementation staff and beneficiaries in using ICTs so that they can easily use mobile app-based application systems. Further, ICT can help us in providing faceless contact reducing transaction time and cost incurred in traditional methods of extension and can address any uncertainty in the minds of the implementation partners/staff/beneficiaries at the time of implementation. Hence, it is vital to employ an innovative ICT system, including cloud, mobile, and geographic information system (GIS) technologies, facilitated by an in-house innovation lab. The ICT component reduces the efforts of smallholding farmers in the long run towards gaining knowledge and support on CNF and SAPs. If a proper ICT support system is established, it can lead to an in-house support system in the agriculture department itself in the long run. Further, ICTs can also help in having real-time monitoring of CNF and SAP practices and proactive data disclosure ensuring data integrity and timely corrective actions for achieving sustainable production.

2. **Raising awareness and financial accessibility for natural farming and SAPs:**
   (a) **Cross-learning and interaction platforms:** It is important to facilitate a mutual understanding of needs, capabilities, and potential areas for synergy. Improving information accessibility is also key; grassroots-level and extension staff should have easy access to market intelligence, best practices, and successful case studies to inform their decisions and strategies, thereby fostering a more informed and cohesive cold storage ecosystem.
   (b) **Learning and exposure visits:** Farmer visits to the neighborhood are very important to gain "learning by seeing." This can be administered through "learning and exposure visits" as well as business management training. These training, exposure visits, and participation of farmers will help them to build knowledge and confidence among farmers to run agri-business (FAO, NSP—Ecological Intensification n.d.-b).
   (c) **Facilitating financial accessibility:** To overcome the issue of capital investments and high interest rates from private investors which are faced by FPOs, it is imperative to establish strong linkages between smallholding farmers and formal financial institutions. By facilitating access to loans with lower interest rates, the member farmers can secure the necessary funds to invest in their FPOs and ensure sustainable growth. Financial constraints

faced by small and marginal farmers and the need for more financial support to adopt natural farming and sustainable agricultural practices are a continuous challenge. The paucity of credit availability to the vulnerable sections on account of their limited asset ownership, lack of formal documents, and mortgageable properties requires support to access institutional finance. By partnering with financial institutions, the project can facilitate easier access to essential resources, such as natural farming inputs, support for SAP practices, seeds, fertilizers, and credit. These measures will not only support farmers in their immediate needs but also promote long-term financial stability and resilience. The linkages with financial institutions addressing the economic barriers can facilitate better adoption and expansion of natural farming and SAP practices having collaboration with institutions for credit provisioning or buyback facilities. Additionally, it is proposed to augment subsidies to promote sustainable production activities among rice growers.

3. **Policy advocacy and carbon mitigation for natural farming and SAPs**
   (a) **Enhancing policy support and advocacy:** Advocacy for supportive policies and regulations is essential to create an enabling environment for strengthening the focus on natural farming and sustainable agricultural practices in rice cultivation. Engaging with policymakers to promote policies that support sustainable agriculture, climate resilience, and farmer welfare can drive long-term change. This includes advocating for policies that provide financial incentives for sustainable practices, ensure fair market access, and support the development of required infrastructure and input supply. Furthermore, promoting the inclusion of climate resilience and sustainability in agricultural education and extension services can ensure that future generations of farmers are well equipped to face the challenges of climate change by adopting sustainable agricultural practices. By enhancing policy support and advocacy on NF and SAPs, we can create a conducive environment for sustainable agricultural development.
   (b) **Need-based agro-advisory services:** To improve farmers' plans for their agricultural activities by adapting to the changing climate conditions, it is recommended that the related stakeholders such as Indian Meteorological Department (IMD) and Krishi Vigyan Kendra (KVK) offer tailored weather forecasts and data related to the management of crops, irrigation, pests, nutrients, soil and water conservation, agribusiness, market, finance, and contingencies through the website, as well as via various information technology (IT) systems like SMS, WhatsApp, mobile apps, and a phone-in help provision. Apart from the generalized location-specific and crop-specific agro-advisories, there is also scope to explore the possibilities to provide customized/need-based and farmer-specific agro-advisory services through different digital modes such as mobile/Web applications, WhatsApp, subject matter experts, and E-consultations using advanced technologies such as artificial intelligence and machine learning which are still in the nascent stage in the agro-advisory sector in India.

(c) **Natural farming and SAPs can facilitate in achieving net zero greenhouse gas emissions:** To align with India's Nationally Determined Contributions (NDCs) and achieve the ambitious goal of net zero, emphasis is required to holistically create awareness among every farmer about carbon emissions and the opportunities within carbon credit markets. By incentivizing sustainable practices, the project can enhance both environmental sustainability and the income of beneficiary farmers. Encouraging farmers to adopt sustainable agricultural practices is crucial. The project should educate them on the benefits of carbon sequestration, which can open up carbon credit markets and provide an additional income stream. Integrating these environmental initiatives into smallholding farmer's production practices will not only promote agricultural sustainability but also contribute significantly to India's broader environmental conservation goals and NDC commitments.

(d) **Fostering community-driven sustainability:** The long-term success of SAPs and natural farming practices relies on a strong focus on community-driven sustainability. Prioritizing community-based natural resource management (NRM) along with individual efforts is crucial for a balanced approach to resource use and sustainable production of rice. A community-level/village-level micro-plan for the development of community-based SAPs in the region is an essential part. Farmers can become self-reliant and climate resilient through the adoption of SAPs and natural farming practices. In the future, any intervention must focus more on these community-based approaches that change the production landscape of the region.

References

Agricultural Statistics at a Glance 2022. Government of India. Ministry of Agriculture & Farmers Welfare, Department of Agriculture & Farmers Welfare. Economics & Statistics Division. Retrieved from, https://agriwelfare.gov.in/Documents/CWWGDATA/Agricultural_Statistics_at_a_Glance_2022_0.pdf

The Fertilizer Association of India- Statistical Database. (2023). ALL INDIA CONSUMPTION OF N, P2O5 and K2O (1950–51 to 2022–23) (n.d.). The Fertilizer Association of India. Retrieved from https://www.faidelhi.org/general/con-npk.pdf

Andhra Pradesh: 1,065 farmers ended lives in 2021, 3rd highe (2022) Retrieved from The Times of India: https://timesofindia.indiatimes.com/city/visakhapatnam/andhra-pradesh-1065-farmers-ended-lives-in-2021-3rd-highest-in-country/articleshow/93846057.cms

Andhra Pradesh - General Scenario and Current Trend (n.d.). Retrieved from https://rkvy.nic.in/static/strategy-of-comendiumandhrapradeshframe.html#:~:text=Average%20land%20holding%20size%20is%201.2%20ha%20and,marginal%20farmers%20having%20less%20than%20one%20hectare%20land

APCNF (n.d.). Retrieved from https://apcnf.in/

Chayal K, Dhaka BL, Poonia MK, Tyagi SVS, Verma SR (2013) Involvement of farm women in decision-making in agriculture. Stud Home Community Sci 7(1):35–37

Duddigan S, Collins CD, Hussain Z, Osbahr H, Shaw LJ, Sinclair F, Sizmur T, Thallam V, Ann Winowiecki L (2022) Impact of zero budget natural farming on crop yields in Andhra Pradesh, SE India. Sustainability 14(3):1689

FAO (2015) International Year of Soil Conference. Retrieved from https://www.fao.org/soils-2015/events/detail/en/c/338738/
FAO (n.d.-a) Changing paradigms of agriculture. Retrieved from https://www.fao.org/agriculture/crops/thematic-itemap/theme/biodiversity/ecological-intensification/zh/#:~:text=Ecological%20intensification%20may%20be%20formally%20defined%20as%20a,improve%20agricultural%20system%20performance%2C%20efficiency%20and%20fa
FAO (n.d.-b) NSP - Ecological Intensification. Retrieved from FAO: https://www.fao.org/agriculture/crops/thematic-sitemap/theme/biodiversity/ecological-intensification/zh/#:~:text=Ecological%20intensification%20may%20be%20formally%20defined%20as%20a,improve%20agricultural%20system%20performance%2C%20efficiency%20and%20fa
Federal Environmental Pesticide Control Act of 1972. (n.d.). Retrieved from https://www.thecre.com/fedlaw/legal14/fedenvp.htm#:~:text=The%201972%20amendments%20established%2C%20under%20the%20Administrator%20of,for%20classifying%20pesticides%20for%20%22general%22%20or%20%22restricted%22%20use
Ghosh N (2004) Reducing dependence on chemical fertilizers and its financial implications for farmers in India. Ecol Econ 49(2):149–162
Gilliom, Robert. J. (2007). Pesticides in U.S. Streams and Groundwater. Environmental science & technology. 41. 3408-14. https://doi.org/10.1021/es072531u
Gulati A, Das S, Khurana M (2024) Zero Budget Natural Farming: Implications for Sustainability, Profitability, and Food Security (No. 24-r-02). Indian Council for Research on International Economic Relations (ICRIER), New Delhi, India
Gupta N, Tripathi S, Dholakia HH (2020) Can zero budget natural farming save input costs and fertiliser subsidies. Report. The Council on Energy, Environment and Water, pp 1–30
John DA, Babu GR (2021) Lessons from the aftermaths of green revolution on food system and health. Front Sustain Food Syst 5:644559
Kumar N, Chhokar RS, Meena RP, Kharub AS, Gill SC, Tripathi SC, Gupta OP, Mangrauthia SK, Sundaram RM, Sawant CP, Gupta A, Naorem A, Kumar M, Singh GP (2021) Challenges and opportunities in productivity and sustainability of rice cultivation system: a critical review in Indian perspective. Cereal Res Commun:1–29
Mohapatra, Souryabrata & Sharp, Basil & Sahoo, Dukhabandhu. (2022). How Changes In Climate Affect Crop Yields In Eastern India. Climate Change Economics. 13. https://doi.org/10.1142/S2010007822500014.
NITI AAYOG - NATURAL FARMING (n.d.). Retrieved from https://naturalfarming.niti.gov.in/andhra-pradesh/
Praveen K (2019) Realizing the potential of a low-cost technology to enhance crop yields: evidence from a meta-analysis of biofertilizers in India
Raeboline, A., Nelson, L. E., Ravichandran, K., and Antony, U. (2019). The impact of the Green Revolution on indigenous crops of India. J Ethn Food. doi: 10.1186/s42779-019-0011-9
Rhodes CJ (2012) Feeding and healing the world: through regenerative agriculture and permaculture. Sage Publications Ltd., London
Ritchie HR-G-O (2024) Population growth. Our World in Data. Retrieved from https://ourworldindata.org/population-growth
Season and Crop Coverage Report, Andhra Pradesh (2023). Retrieved from https://www.apagrisnet.gov.in/2023/weekly/June/weekly_report_(Kharif)_23-08-2023.pdf
Velayutham LK, Damodaran K (2015) An economic analysis of chillies production in Guntur district of Andhra Pradesh. Int J Res Econ Soc Sci 5(9):41–47
World Bank Group (2012) India Groundwater: a Valuable but Diminishing Resource. Retrieved from https://www.worldbank.org/en/news/feature/2012/03/06/india-groundwater-criticaldiminishing#:~:text=India%20is%20the%20largest%20user%20of%20groundwater%20in,of%20drinking%20water%20supplies%20are%20dependent%20on%20groundwater
Xiao, Haibing, Zhongwu Li, Xiaofeng Chang, Bin Huang, Xiaodong Nie, Chun Liu, Lin Liu, Danyang Wang, and Jieyu Jiang. (2018). "The mineralization and sequestration of organic carbon in relation to agricultural soil erosion." Geoderma 329: 73-81.

GPSR Compliance

The European Union's (EU) General Product Safety Regulation (GPSR) is a set of rules that requires consumer products to be safe and our obligations to ensure this.

If you have any concerns about our products, you can contact us on ProductSafety@springernature.com

In case Publisher is established outside the EU, the EU authorized representative is:

Springer Nature Customer Service Center GmbH
Europaplatz 3
69115 Heidelberg, Germany

Batch number: 10703869

Printed by Printforce, the Netherlands